BEE PREPARED
with
HONEY

BEE PREPARED
with
HONEY

Arthur W. Andersen

INTERNATIONAL STANDARD BOOK NUMBER
0-88290-053-6

LIBRARY OF CONGRESS CATALOG CARD NUMBER
75-17102

Third Printing
April, 1977

Printed in the
United States of America
by

Post Office Box 490
50 South 500 West
Bountiful, Utah 84010
295-9451

ACKNOWLEDGMENTS

This book is dedicated to my wife Clara and to my fifteen children, eighty-three grandchildren, seventy great grandchildren, and to my many friends and well-wishers. Without the encouragement of these people this book would not have been written.

Special appreciation should be expressed to the following: my oldest daughter Lyona and her husband, Julian, who assisted me from the beginning, spending many early morning hours typing; my son Mark E. and Dr. Kay J. for their constant help; and to my daughter, LeArta, and her husband, Clark Moulton, for their creative ideas. If this book has detailed simplicity suited for beginners and recognition of the values of honey, I owe it to them.

My son-in-law, Scott Whitaker has given of his time and talent in suggesting manuscript revisions and for drawing the cartoon bees for the cover and chapter illustrations. Another son-in-law, Robert Stum, has taken the photographs and suggested the arrangement of the illustrations. My daughter, Esther, has shared her experience in writing. She and her husband, Russell Dickey, have freely offered their lush garden and lovely home, adding to the incentive and pleasure while writing. My wife, Clara, has given me constant support by both typing the manuscript and assuming full responsibility for home duties in my absence. My thanks is extended also to my granddaughters, Kathi and Gail Andersen, who re-typed the manuscript in looseleaf form. Appreciation is also expressed to my grandson, Douglas A. Whitaker, who has spent many hours editing, improving the script, and indexing. My friend, Roger Hatch, also assisted generously with photography and has shown great interest in the success of this book.

Several recipes have been used from books written by two of my daughters by permission of Bookcraft Publishers, and appreciation is extended to them for that privilege. They are as follows:

Esther Dickey, *Passport to Survival,* (Salt Lake City, Utah: Bookcraft, Inc., 1969.) *Bread Sticks,* page 27; *All-Day Suckers,* page 44; *Basic Candy,* page 45. (Page numbers refer to this book.)

Ruth Laughlin, *Natural Sweets and Treats*, (Salt Lake City, Utah: Bookcraft, Inc., 1973.) *Honey Lemon Bread*, page 27; *Honey Sponge Cake*, page 38; *Honey Marshmallows*, page 44; *Pumpkin Pecan Pie*, page 52.

The following recipes were contributed by my daughter LeArta Moulton: *Fruit Drinks*, page 23; *Hot Carob Drink*, page 24; *Delicious Whole Wheat Bread*, page 26; *Honey Baked Parsnips*, page 34; *Carrots with Lemon, Butter & Honey*, page 34; *Filled Cookies*, page 50; *Millet Pudding.* page 56; *Vanilla Ice Cream*, page 62; *Ice Cream Tiger*, page 62; *Carob Ice Cream*, page 62; *Raspberry & Strawberry Ice Cream*, page 62; from *The Gluten Book* (LeArta Moulton). *Horehound coughdrops*, page 73; *Iced Herb Tea*, page 73; from *Nature's Medicine Chest* (LeArta Moulton).

Appreciation is also extended to Rev. Maurice H. Wess, 1523 19th St., Denver Colorado, 80202, for his contributions to the recipe section taken from his book, *Honey, I Love You.*

PREFACE

Honey has been considered one of man's most delectable and nutritious foods, from earliest times. More and more people are considering it more than an occasional confection. They are using it in cooking, canning, and in almost every other way in which sugar is used. Evidence is increasing that there are significant health values in the moderate consumption of this natural sweet.

Numerous requests from grandchildren and other family members and friends led me to write this book as a guide to back-yard beekeeping and the value of honey. As the demand for this natural sweet increases and the supply gets smaller, prices will soar to the point where we will be deprived of one of our best foods unless we learn to provide for ourselves.

Much of the material for this book will be taken from my eighty years of personal experience in commercial beekeeping. This publication completes the fulfillment of my daydream as an eight-year-old lad who stood with playmates by the horse-driven wagon of John Zwahlen. From a 40-gallon barrel of beautiful hard white honey, he was deftly spading out huge chunks with a large hardwood paddle and selling it by the pound. As we watched the glistening white sweetness fall into a pan provided by my mother, our covetous looks tempted this kindly man to reprimand us. But then he smiled and turned the heavy laden spoon toward us, depositing a generous helping as we held out our cupped, unwashed hands. After that feast I told Mother, "When I get big, I am going to get some bees and have all the honey I can eat." This was the resolve of a sweets-hungry eight-year-old. The fulfillment of this longing came many years later with my possession of 800 colonies and a bumper corp of 36,000 pounds of beautiful white honey.

Ecologists should welcome a book that describes the bee as a vital force in maintaining nature's balance. Those concerned with home food storage should also find this book valuable. One or more hives of bees, kept in the garden, the orchard, or farm, can be considered a food storage item which also supplies daily demands. Honey, like other basic foods, loses some value over long periods of storage. The delicacy and freshness of new honey with all the enzymes, digestive properties, and many known food values can be a prized possession.

It is hoped that *Bee Prepared With Honey* will satisfy the needs of those interested in beekeeping as a hobby as well as those willing to go a little further into a small business venture. It is also hoped that educators will find it valuable as a practical guide for studying the bee and the production of honey and may consider the possibility of maintaining a few hives at one end of the school yard for natural science projects.

To my family, to neighbors, to members of scout troops, to students, to hobbyists, and to all who may be interested in learning the art of neighborhood beekeeping, I extend my welcome and a promise. When you open the hive and get the first glimpse into the miraculous behavior of honey bees and the lessons they teach, your initial thrill will be increased a hundred fold as you continue your search.

As I approach the age of 93, I still find pleasure in sitting by a beehive watching a stream of "take-offs" and "landings." Some of the bees are laden with pollen, others have bellies rounded with filled honey sacks, the nectar soon to be transformed into sweetness. Only God in His great wisdom could have created the magic of the hive!

Arthur W. Andersen

TABLE OF CONTENTS

   Honey Butter; A Morning Lift; Fruit Hot Punch; Orange
   Julius; Prize Lemonade; Marmalade; Banana Slush; Fresh
   Fruit Slush; Apple Drink; Lemonade Slush; Honeyed Fruit
   Juice; Fruit Punch; Party Punch; Orangeade; Honey Pick-
   Up, Fruit Drinks: Basic Recipe; Honey on Grapefruit;
   Fresh Fruit Cup; Hot Carob Drink; Fruit Punch Concen-
   trate; Honey-Orange Sauce; Honey Root Beer; Banana Egg
   Nog.

   Delicious Whole Wheat Bread; Bread Sticks; Honey Lemon
   Bread; Honey Can Bread; Honey Fruit Breads; Apple Wal-
   nut Loaf; Honey Rye Bread (round); Date Nut Bread;
   Norsk Holiday Bread; Zucchini Bread; Honey-Prune Bread;
   Honeybutter Sauce; Honey Sandwich Filling; Honey Salad
   Dressing; Sandwich Filling; Other Suggested Sandwich
   Fillings; Honey Butter; Fruit Salad Dressing; Fluffy Cottage
   Cheese Dressing.

   Cabbage, Sweet and Sour; Honeyed Sweet Potatoes; Beets
   in Yogurt; Sweet Sour Lentil Sauce; Candied Yams; Apple
   Yam Casserole; Carrots with Lemon, Butter and Honey;
   Honey Baked Parsnips; Beets; Cabbage Slaw; Crystal Carrots;
   Rotkohl (Red Cabbage).

CHAPTER 1

THE NUTRITIONAL VALUES OF HONEY

Honey Is A Valuable Addition To The Diet

There is confusion surrounding the value and properties of honey. Some have claimed that it is no better or worse than refined sugar; others have advanced extravagant claims for the value of honey. As usual, the truth lies somewhere in between. At this juncture it would seem proper to consider documentation from the American Medical Association:

> Honey is a natural, unrefined, nutritionally valuable food. It is unique in that it is the only unmanufactured sweet available in commercial quantities.
>
> Seventy-five per cent of its composition is sugars which give honey an energy producing value second to few foods. Commonly used sugars, cane and beet, must be broken down into simpler sugars by digestive juices before they can be assimilated into the blood stream and tissues. These resulting simple sugars, dextrose and levulose, make up almost the entire sugar content of honey. This makes it reasonable to conclude that little digestion is necessary and absorption takes place quickly. It is also reasonable to conclude that, except in unusual cases, the utilization of honey puts no tax upon the digestive system.
>
> In addition to the sugar content of honey, it contains a small quantity of various minerals necessary for good nutrition and undetermined constituents.[1]

Honey Is Useful For Feeding Infants And Children

Ask your physician for a baby formula in which honey is included. Because of the composition of honey, it is recognized as a supplement to milk in infant feeding. Since it is easily obtained and is very palatable and digestible, honey is a form of carbohydrate which should have wider use in infant feeding. Honey is also an excellent source of readily-available energy for growing children. It is good for all ages.

In the last few years, Dr. Schultz and Dr. Kraft of the Chicago University, have done research work which has shown the superiority of honey over carbohydrates for infant feeding. This work

---

1. *Starting Right With Bees*, A. I. Root Co., page 76. *Starting Right With Bees* was first written in 1944 by H. G. Howe and later revised and edited by Editorial Staff of *Gleanings in Bee Culture* under direction of John Root, Seventeenth edition being published in 1974 by A. I. Root Company, of Modena, Ohio.

was recorded in the *Journal of Pediatrics*.[2] It was also reviewed in the *Journal of the AMA*.[3] We quote from the last named publication by permission:

> It was found that honey is easily digested, being absorbed more rapidly than most sugars during the first 15 minutes after ingestion. It is well tolerated by the infant and does not cause diarrhea. It may facilitate weight gains, since there was a slight tendency for greater average gains daily to occur on lower calorie intakes when honey was included in the formula. It appears that honey should be recommended for wider use in diets of infants.

Levulose and dextrose, the two principal constituents in honey, are both blood sugars. The combination of the two make what is known as invert sugar. When honey is eaten, it is almost immediately inverted into the blood stream. Scientists have shown that the levulose, which is sweeter than ordinary sugar, disappears and apparently becomes dextrose, a sugar found in ordinary blood. Apparently nature has designed that the two sugars should go together and, according to some of the best medical authority, these two sugars in honey make an ideal sweet.

Drs. Schultz and Knott have another article in the *Journal of Pediatrics*[4] entitled, "The Effect of Honey upon Calcium Retention." From the general summary of this article we quote by permission:

> The data have been divided to show the effects of honey in contrast to corn syrup....The same infants were fed under comparable conditions to make the comparison of these two types of carbohydrates completely valid....The average retention of calcium was always higher if honey has been included in the fromula. The increased retentions where honey was fed occurred both with low and with high intakes of vitamin D and occurred regardless of the type of milk fed in the use of lactic acid. Honey, therefore, would seem to have a definite influence on the retention of calcium by young infants.
>
> The uniform improved response with honey is of special interest in view of the validity of results observed. While some of them had higher results on corn syrup when paired periods were compared, the majority of the infants had better results on honey and the averages for the paired periods were consistently in favor of honey.

---

2. June, October, November, 1938.
3. November, 1938.
4. October, 1941.

Honey Aids Digestion

Honey has been proved to be one of the natural foods high in Vitamins, especially vitamin B1. It is now evident that men of high standing in the medical profession, both in this country and Europe, recognize honey as a superior sweet for children, old people and those of impaired assimilation. If cane or beet sugar remains in the alimentary canal too long, fermentation may take place, causing gas and distress and what is sometimes called old-fashioned heartburn. In contrast, honey, is absorbed immediately and no distress occurs.

Use Honey Instead Of Sugar

There are numerous myths about the difficulty of substituting honey for sugar in cooking. Most of these have proved to be false. With some experience and attention to the numerous honey recipes now available there is scarcely a food which calls for sugar that cannot be prepared with honey.

Only about half as much honey is required in recipes calling for sugar, so even with the inflated price of honey, the cost differential is narrowed.

Add small amounts to cooked vegetables such as peas, corn and string beans to bring out flavor. Substitute it wholly or in part for sugar in ice cream, salad dressing, bread, cakes, puddings and gelatins. Add it to whipped cream instead of sugar. Use it on cereals. Use a mixture of honey and cream cheese for a sandwich filling. Use it on baked apples. Mix it with crushed fruit for short cakes. Use it in fruit drinks, for sundaes, and for topping ice cream along with fruit or nuts.

The author has been the beneficiary of honey over the years, counting it a perfect food. It builds, it heals, it endures. For example, the author has in his possession a honeycomb built 52 years ago by the bees in shape to resemble a bust of Uncle Sam. It has darkened in color and has lost some of its nutritional values but is is still sweet to the taste.

CHAPTER 2

TIPS FOR USING HONEY

Dilute For Pouring: Pure or raw honey can be diluted with water up to 20% without danger of fermentation. This mixture is convenient for general use in honey dispensers, hot cakes, cereals, etc.

Restoring To Liquid Form: Hard or granulated honey is pure ripened honey and is not spoiled. Restore it to a liquid form by placing the container in a pan of warm water, providing space under the container by setting it on widemouth bottle rings. Keep. the heat to a low temperature. Honey subjected to excessive high heat destroys the enzymes and other digestive properties, thus rendering it of little more value than any other sweetener.

Measuring: To measure honey with ease, wet or oil the cup or spoon first, or use the same cup used in measuring oil or melted butter.

Substitute For Other Sweeteners: Honey may be used in place of other sweeteners. Replace equal amounts up to one cup.

Allowing For Natural Moisture: Decrease liquid ¼ cup for each cup of honey used to adjust for natural moisture found in honey.

Adjusting Baking Temperature: Reduce baking temperature 25 degrees to prevent over-browning.

Age Increases Flavor: Honey-sweetened baked goods are more flavorful the next day. Honey cakes and cookies are tender and moist

Adding To Shortening: Add honey to shortening slowly, in a fine stream, blending well, for better volume and texture.

Cakes: In cakes, honey can replace as much as one-half of the sugar without making it necessary to change the proportions of the other ingredients in the recipe.

Cookies: In cookies, the amount of honey that can replace the sugar varies with the type of cookie. For example, replace no more than 1/3 of the sugar with honey. For brownies, honey may be used for half of the sugar called for. For fruit bars, honey can be used for up to 2/3 of the sugar.

Thorough Mixing: In both cakes and cookies, the honey should be mixed thoroughly with the other ingredients to prevent a soggy layer from being formed on top. Combine honey with either the shortening or the liquid.

Choosing Honey Flavors: Use mild-flavored honey for bland dishes. Honeys with stronger flavor may be used in spice cake, gingerbread, brownies, and in other rather highly flavored products.

Empty Containers Completely: In any recipe, be sure to scrape all the honey out of the measure for accurate amounts.

Use Honey Recipes: For cakes and cookies made with honey, you'll get best results if you use recipes developed especially with honey as an ingredient. However, you can use honey for part of the sugar in other recipes as suggested above.

Carob: Carob is a food with high nutritious value which resembles cocoa and chocolate in flavor and texture. It contains thiamin, riboflavin, Vitamin A and also important minerals: calcium, phosophous, iron, copper, magnesium.

It has none of the allergic properties chocolate has and is much lower in calories and is not a stimulant. It has a high pectin content. It offers a natural protection against diarrhea. Carob is available in powder form, chunks or bars. It may be melted and used like cooking chocolate, can be cut in small bits for carob-like chocolate chips.

Serving Honey Uncooked: Honey is at its best uncooked. Try it in the following ways to accent or enhance flavor of other foods.

Use As A Spread: Use honey with bread, toast, pancakes, waffles, or other hot breads. Serve it "as is," or mix it with equal parts of butter or margarine for a honey spread.

Or, mix equal parts of honey, peanut butter, and butter or margarine.

Or, make Swiss honey by mixing equal parts honey, butter or margarine and whipped cream.

Use As A Topping: Make a simple topping for fresh fruit, fruit salad, fruit cup, or ice cream by drizzling the food with honey just before serving. To do this, hold a container of warmed honey a short distance above the food and let the honey drop in a fine stream.

Whipped cream sweetened with honey makes a good topping for plain or fruit ice cream or gelatin desserts.

On hot desserts, try honey hard sauce. To make it, cream honey with one-half as much butter or margarine until well blended.

Use As A Sweetener: Serve honey on hot or cold cereals.

In beverages, dilute honey by mixing equal parts of honey and water. Use it with grape juice, in lemonade or orangeade, hot spiced tea, with milk, and in eggnogs.

PART I

HONEY RECIPES

CHAPTER 3

"NON-COOKING" HONEY RECIPES

Honey Butter

1 cup honey   ¾ cup butter or margarine

Let the butter or margarine stand at room temperature until soft enough to blend with the honey. Stir until thoroughly mixed or put it in a blender or electric beater until fluffy. Pack in a wide-mouthed jar. Cover tightly and store overnight in the refrigerator to solidify. Spread like butter on hot biscuits, waffles, or pancakes.
Variation: add 1 egg yolk and 4 drops of vanilla, beat.

A Morning Lift

2 oranges) Peel, leave the white   2 Tbsp honey
1 lemon) for bioflavonoids   1 cup water
2 apples   (peeling & cores)   1 Tbsp protein powder
6 dates   or nuts

If you have a high speed blender, blend for 3 minutes. Makes 4 servings.
Variations: Substitute apples with 1 cup apple juice or 2 Tbsp of apple concentrate and 1 cup water or pineapple, grape, grapefruit, banana, tomato, apricot, avocado, grapes.

Fruit Hot Punch

¾ cup honey   3 lemons
1 cup sugar   1 6-oz. can frozen orange
2 quarts water   juice
1 tsp almond

Boil honey, sugar, water a few minutes; then add other ingredients and simmer about 5 minutes. Serves 20.

Orange Julius

1 6-oz. can orange juice   1 egg
   concentrate   1 Tbsp honey

1 cup water
1 cup milk

1 tsp vanilla
6 ice cubes

Beat in blender until ice cubes are dissolved. Serves 3.

Prize Lemonade

Cut in pieces:
1 lemon — cut off ends and use remaining rind and pulp, and
1/4 orange — rind and all.

Blend with 2 cups water for 3 minutes. Add water to make up to 1 quart. Strain. Sweeten with 1/3 cup honey (warm 1/2 cup juice and mix honey in). Cool with ice.

Marmalade

Combine strained pulp with equal amount of honey. Cook for 5 minutes. Cool. GOOD.

Banana Slush

1 cup water
¾ cup honey
juice of 1 orange

juice of 1 lemon
2 bananas, mashed

Add hot water to honey and dissolve. Add remaining ingredients and freeze. When ready to use, spoon frozen mixture into cups. Add ginger ale and garnish with cherry.

Fresh Fruit Slush

Sweeten with honey any ripe fruit that has been pureed — freeze and use with orange juice or lemonade as a drink.

Apple Drink

3 cups orange juice
2 unpeeled apples (cut up)
1 avocado

3 Tbsp honey
1/3 cup cashew nuts

Mix in a blender. This makes about 4 cups of juice.

Lemonade Slush

1-12 oz. can of frozen lemonade	8 cans of water
	3 cups of honey

Heat until the honey dissolves, and freeze. Take out 2 hours before serving, mash and add 7-Up.

Honeyed Fruit Juice

1 cup orange juice	2 cups water
1 cup lemon juice	4 Tbsp honey
2 cups grape juice	

Mix thoroughly, chill. Serve cold. Makes 6 servings.

Fruit Punch

1 cup berries in season	½ cup lemon juice
1 cup honey (mild)	1 quart mint tea
1½ cups orange juice	

Crush berries and combine all ingredients. Chill. When ready to serve, add crushed ice. Makes 20 servings.

Party Punch

2 cups honey	1 cup peppermint tea leaves
1 cup hot water	3 quarts water
1 cup lemon juice	4 quarts of ginger ale
3 cups orange juice	1 cup maraschino cherries

Add honey to hot water. Mix well. Add fruit juices, mint leaves and water. Chill thoroughly. Add ginger ale and cherries just before serving. Pour over ice cubes in punch bowl. Makes 40 servings.

Orangeade

2 cups orange juice	½ cup honey (mild)
½ cup lemon juice	1 cup water

Combine ingredients and stir well to dissolve honey. Pour

over cracked ice in tall glasses. Garnish with orange slices, mint and cherries.

Honey Pick-Up

1½ cups fruit juice   ¾ Tbsp honey

Mix for only 10 seconds.

Fruit Drinks: Basic Recipe

3 cups frozen orange juice   ½ cup fresh or frozen
1 tsp honey   fruit

(When using fresh fruit, add enough ice to make it the thickness desired.) Add ingredients to blender and mix 30 seconds or until smooth.

1. Avocado and Orange Juice

1 Avocado
1 tsp cinnamon
1 tsp honey
Follow basic recipe.

2. Strawberry and Orange Juice

2 cups frozen strawberries
2 cups orange juice
1 tsp honey
Follow basic recipe.

3. Watercress or Comfrey and Pineapple

2 cups canned pineapple
  juice (fresh if available)
½ tsp honey
1 cup chopped watercress
  or comfrey

Blend in liquifier (strain leaf bits out if not desired) Blend all ingredients 30 seconds. Serve immediately.

4. Carob and Milk

2 cups milk
2 tsp carob powder
2 tsp protein powder
1 Tbsp honey

Mix in blender. If thicker consistency is desired, add ice.

Honey on Grapefruit

Cut a grapefruit in half. (Serrate if desired) Add a tablespoon of honey spread over the top and place in the refrigerator overnight before serving. During this time, the honey dissolves into the juice of the grapefruit with a pleasing tone, being neither sweet nor sour.

Variation: Prepare grapefruit as above, then place in oven on low heat, 200 degrees for about 20 minutes. Serve while hot.

Fresh Fruit Cup

2½ cups orange sections
1 cup banana slices
1 cup unpeeled redskinned
   apple slices

1 cup orange juice
Dash of lemon juice
Honey to taste

Combine and serve, chilled, as appetizer or dessert. Garnish with cherries. Makes 6 servings.

Hot Carob Drink

3 cups milk (or soymilk)
2 Tbsp carob powder

2 Tbsp honey

Heat milk suitable for drinking, add the carob and honey.

Fruit Punch Concentrate

4 cans frozen orange juice
1-46 oz. can pineapple juice
2 pkgs cherry Koolaid
4 pkgs orange Koolaid

5 bananas, blended
2 cups honey dissolved in
   1 gallon of warm water
2½ cups sugar

Blend the above well, refrigerate. Will keep well for several weeks. 2-1/2 parts water added to 1 part concentrate. (Good color, excellent flavor)

Honey-Orange Sauce

½ cup liquid honey
¼ cup orange juice

1 tsp grated orange rind
Few grains of salt

Mix ingredients until well blended. Makes ¾ cup. Serve on pancakes or waffles.

Honey Root Beer

2 Tbsp hops (optional)
1 yeast cake or 1 Tbsp of
   dry yeast
4 gallons of warm water

1 bottle of Hires Root
   Beer Extract
6 cups honey
Dry ice (optional)

Bring 2 cups of water to a boil, turn heat off and add the 2 Tbsp of hops. While this is steeping for 20 minutes, add the yeast to 1/2 cup of warm water with 1 tsp sugar in, and let it work. Have a large container (a canner is good) and measure 4 gallons of warm water. Dissolve the honey, root beer extract, the strained hops and worked yeast, and mix well. Let all work together until a foam comes on top, then bottle in 2-quart or gallon bottles. It can store in the refrigerator for a few days if you will be using it up by then. If not, it may be corked in pop bottles filled to within 1/2 inch at top (more air space may cause spoilage) and store in a cool place. If a party is to be served the next day, add dry ice in the container, and serve from there.

Banana Egg Nog

2 cups milk	1 egg
3 ripe bananas	nutmeg or cinnamon
2 to 3 Tbsp honey	

Combine all ingredients except spice in blender or electric mixer bowl. Blend or mix at high speed until egg nog is smooth and frothy. Pour into 4 glasses; top with spice and serve.

Variations: strawberries or any fruit in season.

CHAPTER 4

BREADS, BUTTERS, AND SANDWICH FILLINGS

Delicious Whole Wheat Bread

This recipe gives you a beautiful texture—the kind you have always wanted to get with your whole wheat flour. It is moist, light, and does not crumble.

6 cups very warm water
3 to 4 Tbsp Lecithin (pwd, granules, or oil)
15½ cups whole wheat flour
2/3 cup honey
2 Tbsp salt
2/3 cup oil

2 Tbsp yeast dissolved in ½ cup warm water plus 1 tsp honey or sugar
1 cup pwd milk—non-instant
1 cup pwd whey or 6 cups liquid whey in place of the 6 cups water above—optional

Sprinkle yeast on honeyed water and set aside to work (Do not stir yeast and water). When yeast has worked up to equal 1 cup it is ready to put in the dough mixture.

Mix: Water and Lecithin, Add 8 cups flour—mix
Add oil, honey and salt—mix
Add worked yeast—mix
Add pwd milk and whey—mix
Add remaining flour—mix (can substitute some of the whole wheat with white)
Knead: 10 min., low speed, with electric bread mixer
Place in pans.
15 min. by hand. Let rise once before putting in pans.
Rise 1/3 in bulk.
Bakes 50 min. (In high altitudes bake at 400 degrees F. for 10 min. then 350 degrees F. for 35 min.)
Makes 4 to 5 loaves

Hint: Flour which has been ground at least 2 hrs. or carried around a while, measures more than the freshly ground flour. For instance: If the recipe calls for 12 cups and you were using freshly ground flour—you could use almost 13 cups.

Bread Sticks

6 cups liquid milk	1/3 cup oil
½ cup honey	2 Tbsp dry yeast
2 Tbsp salt	12 cups flour

Dissolve yeast in ¼ cup warm water and 1 tsp honey. Mix with liquid milk, oil, salt and rest of honey. Stir in about half the flour. Put portions of dough in nest of flour on the board, ¼ of dough at a time. Flour the hands. Using as little flour as possible, shape dough into four round balls that are not sticky.

Combine the flour balls and develop gluten for about 10 minutes until the dough is springy, smooth, elastic and easy to handle. Put the ball of dough smooth side up in a warm, lightly-greased bowl and keep it warm until dough doubles in size.

On bread board roll out dough ½ inch thick. Cut with circular cookie cutter and snip each circle into two equal pieces with scissors. Roll each of these pieces into an oblong. Dip in egg yolk or canned milk and roll in sesame seed. Allow to rise until dough again doubles in bulk, then bake at 400 degrees F. until brown and crisp. Makes about 100 bread sticks.

Variations: (1) Bread stick dough can be shaped into leaves. Make fancy bread by braiding three ropes of dough; or snip top of loaf with scissors in two rows down length of loaf. Sprinkle with sesame seed. Make raisin bread or flavor with onion or garlic. (2) Use same dough to make hamburger buns, muffins or sweet rolls. (3) For richer dough and a more tender product add 2 eggs to recipe and increase oil to 2/3 cup and honey to 3/4 cup.

SWEET BREADS

Honey Lemon Bread

6 Tbsp margarine	½ tsp soda
2 eggs	1 tsp (rounded) baking pwd
1 cup honey	½ tsp salt
1 cup milk	½ cup nuts
2 cups whole wheat flour	grated rind of one lemon

Cream butter and honey. Add eggs and beat well. Add milk and dry ingredients alternately. Add lemon rind and nuts. Bake in loaf tin 325 degrees F. for 1 hour.

Topping: Mix juice of 1 lemon with ½ cup warm honey and spoon over hot loaf while still in pan.

Honey Can Bread

3¾ cups raisins	2 eggs
1½ cups boiling water	1¼ cups honey
1½ tsp soda	1 tsp vanilla
4 Tbsp butter	4 cups whole wheat flour
½ cup brown sugar	1 tsp salt

Add boiling water to raisins and set aside to cool. Add soda. Cream butter, honey, eggs, add honey and vanilla. Add whole wheat flour and salt with raisin mixture to above. Beat well. Bake in tall soup cans, greased and half filled, at 350 degrees F. for 45 minutes.

Honey Fruit Breads

1½ pkg. or 1½ Tbsp dry yeast dissolved in ½ cup warm water	¾ cup honey
	1 Tbsp salt
1 egg, beaten	5 Tbsp cooking oil
1 cup milk	7 cups whole wheat flour, (approximately)
1½ cups hot water	

Mix together in large bowl egg, milk, honey, salt, hot water, and oil. Add 3 cups flour and beat until smooth. Add yeast mixture and stir in. Add flour to form a soft dough that is easy to handle. Knead 10 minutes. Form into a ball and let stand 20 minutes for easy handling. At this stage add fruits.

Variations: The following can be added to each loaf. Add:

1 cup finely chopped raw apples	1 tsp cinnamon
	1 cup chopped walnuts

or:

Any softened dried fruit can be used.

Apple Walnut Loaf

1½ cups sifted enriched flour	1 cup broken walnut meats
2 tsp baking powder	¾ cup chopped apples
½ tsp baking soda	1 egg slightly beaten
1 tsp salt	¼ cup brown sugar
1 tsp cinnamon	¼ cup honey
1/8 tsp allspice	1¼ cups buttermilk
1½ cups crushed ready-to-serve wheat cereal flakes	2 Tbsp melted shortening
	¼ tsp nutmeg

Set oven for moderate, 350 degrees F. Mix and sift flour, baking powder, baking soda, salt and spices. Add cereal flakes, walnuts and apple. Combine egg, brown sugar, honey, buttermilk and shortening; add, mix just enough to moisten dry ingredients. Do not beat. Turn into greased loaf pan 8 x 3 x 4 inches. Bake 1 hour. Makes 1 loaf

Honey Rye Bread (round)

2 cups Rye flour	1½ tsp annise
3 cups whole wheat flour	2½ tsp caraway
1 cup white flour	2 cups water
2 tsp salt	2 Tbsp lecithin
½ cup honey	2 Tbsp yeast
½ cup molasses	½ cup warm water for
5 Tbsp oil	yeast to work in

Add sufficient flour and rye flour to make a soft dough. Let raise to double in bulk, knead and form into round loaves on a cookie sheet. Bake at 325 degrees F. for 1 hour, or until done.

Date Nut Bread

1 cup boiling water	1 egg
1 cup chopped dates	1½ cups unbleached flour
2 Tbsp lecithin	¾ tsp salt
2 Tbsp shortening	2 tsp alum-free
¾ cup honey	baking powder
1 cup broken nut meat	

Add boiling water to chopped dates and cook about two minutes, stirring constantly. Mix egg, shortening and lecithin and beat. Add sifted dry ingredients, nuts and honey. Mix well. Pour mixture into greased pan.

Bake at 325 degrees F. for one hour and 15 min. or until done. One loaf.

Norsk Holiday Bread

1 package muffin mix	½ cup chopped
1¼ cup honey	candied cherries
1 cup golden raisins	

Follow easy direction on muffin mix package. Add honey and fruit, mixing thoroughly. Let rise. Knead on floured surface until smooth. Divide dough in half and shape into 2 round loaves in 8" layer pan. Punch center to give typical shape. Double in size and bake at 350 degrees F. for 30 min. Makes two loaves.

Zucchini Bread

3 eggs	2 tsp soda
¾ cup oil	1 tsp salt
2 cups honey	¼ tsp baking powder
2 cups zucchini--grated	3 tsp cinnamon
3 tsp vanilla	1 cup nuts
3 cups flour	

Mix all ingredients. Bake at 325 degrees F. for 60 to 70 min. or until done.

Honey-Prune Bread

2 Tbsp shortening	½ tsp salt
1/3 cup liquid honey	½ tsp soda
3 Tbsp sugar	½ cup sour milk
2 eggs	½ Cup chopped
1 cup sifted all-purpose flour	cooked prunes
2/3 cup whole-wheat flour	½ cup chopped pecans
1½ tsp baking powder	½ tsp grated lemon rind

Cream shortening, honey and sugar thoroughly. Add eggs; beat until well mixed. Sift together flour, whole-wheat flour, baking powder, salt and soda. Add dry ingredients and sour milk alternately to the creamed mixture and beat until well blended. Stir in prunes, pecans and lemon rind.

Pour into greased loaf pan, 8½ x 4½ x 2½ inches. Bake at 350 degrees F. (moderate oven) for 1 hour and 10 minutes. Makes 1 loaf, about thirty-six ½ inch slices.

Variations:

Apples, apricots or other fruits could be used.

Honeybutter Sauce

Warm 1 cup liquid honey in double boiler: add ¼ cup butter or margarine, ¼ tsp cinnamon and dash of nutmeg. Serve warm. Use on Waffles, pancakes, etc.

Honey Sandwich Filling

3 Tbsp honey   4 oz. cream cheese
  (liquid or creamed)

Beat honey and cream cheese together until light and fluffy. For toast topping: Add 4 tsp orange juice, 1 tsp orange rind, 1 tsp pineapple juice (optional) to the sandwich filling. Top with sliced almonds. Makes 1 cup enough for about 8 sandwiches.

Honey Salad Dressing

1 tsp paprika	½ cup honey
½ tsp powdered dry mustard	3 Tbsp lemon juice
½ tsp salt	¼ cup vinegar
½ tsp celery salt	1 cup salad oil

Mix the dry ingredients. Add the honey, lemon juice, and vinegar. Slowly add the salad oil, beating until well blended. Makes about 2 cups.

Sandwich Filling

2 avocados	dash of garlic powder
1 Tbsp lemon juice	1 tsp dill weed or
1 tomato	other herbs
1 Tbsp honey	1 finely minced green onion
1 Tbsp of mayonnaise	

Mash avocados and tomatoes together and add the other ingredients. This dip is very good with celery sticks, or refrigerated for stuffed celery. Could be used for sandwich filling.

Other Suggested Sandwich Fillings

Honey with chopped dried fruits and chopped nuts if desired.
Honey and chopped or grated orange peel
Honey and peanut or almond butter

Honey spread with chopped nuts or grated orange peel.

Honey Butter

One cube butter or ½ cup   2 cups honey
1 egg yolk   ½ tsp vanilla

Beat until fluffy. Keep refrigerated.
Variations: Add grated orange rind, crushed pineapple, cinnamon or nutmeg or ½ tsp coconut flavoring and ½ cup coconut.

Fruit Salad Dressing

3 eggs   2 Tbsp orange con-
½ cup honey   centrate
   or part sugar   1 tsp grated lemon peel
   juice from one lemon
1/8 tsp salt

Beat and thicken over in double boiler. Cool, add ½ pt. cream whipped. Serve with pears, peaches, mandcrin oranges, grapes, canned or fresh bananas, pineapple chunks, marshmallows. Serves approximately 30.

Fluffy Cottage Cheese Dressing

2 cups cottage cheese   2 Tbsp orange juice
1 Tbsp lemon   1 Tbsp honey

Blend all ingredients in electric beater or blender.

CHAPTER 5

VEGETABLES

Cabbage, Sweet and Sour

4 cups red or green cabbage shredded	½ cup sweet cider
3 onions, grated	3 Tbsp honey
Juice of 2 lemons	2 Tbsp oil
4 Tart apples with skins, (diced)	1 Tbsp caraway seeds
½ cup raisins	pinch of allspice (ground)

Blend all ingredients in saucepan. Cover, simmer gently for 10 minutes.

Honeyed Sweet Potatoes

6 sweet potatoes or yams, cooked	Juice of 1 lemon
pinch of ground allspice	¼ cup oil
	½ cup honey

Arrange sliced sweet potatoes in an oiled casserole dish. Mix remaining ingredients together. Pour mixture over potatoes. Bake at 350 degrees F. for about 30 min., basting occasionally with liquid. Serves 6.

Beets in Yogurt

3 cups beets, grated	1/8 basil
beet tops, minced	¼ cup stock
pinch of salt	1 tsp honey
1 bay leaf	¼ cup yogurt

Place beets, tops, salt, bay leaf, basil and stock in steamer. Cover and steam gently for 5 minutes. Remove bay leaf. Blend in honey and yogurt. Serves 6.

Sweet Sour Lentil Sauce

2 cups lentils, cooked	3 Tbsp cider vinegar
3 Tbsp honey	3 Tbsp parsley, minced

Combine all ingredients, simmer for 10 minutes in covered saucepan. Good with plain steamed fish. Serves 6.

Candied Yams

Slice cooked yams, brown lightly in butter. Drizzle honey over yams and let it melt in. Sprinkle lightly with salt and pepper and serve hot. Parsnips can be used in place of yams.

Apple Yam Casserole

6-7 yams   5-6 cooking apples

Par boil yams 20 minutes; peel, slice apples. Layer with sauce. Bake 1 hour at 350 degrees F.

Sauce:

¾ cup honey   2 cups fruit juice (orange)
2 Tbsp corn starch   1 tsp salt
2 tsp lemon juice

Mix. Cook until thick.

Carrots with Lemon, Butter and Honey

3 cups chopped carrots   1 tsp honey
  (cubed or sliced)   salt to taste
2 Tbsp butter or substitute   corn starch-optional
2 tsp lemon juice

Simmer carrots until just tender. Add remaining ingredients (thicken with corn starch if desired). Simmer lightly another 10 minutes.

Honey Baked Parsnips

Cook parsnips—steam or boil with peelings on.
Make a layer of thinly-sliced parsnips in baking dish. Dot generously with butter and drizzle warm honey around in circles. Salt to taste. Repeat layers. Bake at 350 degrees F. until heated through, (about 25 minutes).

Beets

6 beets	½ cup yogurt
½ cup fresh lemon juice	1 tsp olive oil
1 tsp cinnamon	1 tsp honey

Blend all ingredients except yogurt in blender until almost totally liquified. Add yogurt, mix a few seconds, remove and chill before serving.

Cabbage Slaw

2 cups grated cabbage	2 Tbsp lemon juice or
2 Tbsp honey	vinegar
2 Tbsp mayonnaise	

Mix all ingredients together and serve on lettuce leaf, with dash of paprika and a sprig of parsley on top of mayonnaise.

Variation: Add a few radishes, tomato, cucumber or even some red cabbage grated in the white cabbage.

Crystal Carrots

1 Tbsp vegetable oil	2 Tbsp honey
2 Tbsp butter or margarine	1 Tbsp lemon juice
12 medium size carrots, cut on	1/3 cup water
bias into ½" thick slices	2 Tbsp chopped parsley
½ tsp salt	

Heat the oil, water, and salt in a skillet. Add carrots and cook while stirring until carrots are just tender. Add the honey, lemon juice and 2 Tbsp margarine. Simmer uncovered until honey is melted and the remaining liquid evaporates and carrots are glazed. Sprinkle with parsley. Serves 6.

Rotkohl (Red Cabbage)

Finely shred one head of red cabbage; add 1 Tbsp oil, 1/3 cup vinegar, 3 Tbsp honey, a pinch on the tip of a knife of ground cloves and ground allspice, salt to taste. Add 1 cup water. Cook in pressure cooker for only three minutes after whistle blows. Cool immediately. Thicken with about 1-2 Tbsp cornstarch. Serve 8-10.

Hawaiian Ham

2 Tbsp butter or margarine	1 tsp curry powder
1 cup diced celery	½ cup chopped onion
1 green pepper cut in thin strips	1 Tbsp cornstarch
	½ cup orange juice
1 can (9 ounces) pineapple tidbits	2 cups cooked ham, cut in thin strips
¼ cup slivered blanched almonds	1 Tbsp honey

Melt butter or margarine in frying pan; add curry powder, celery, onion and green pepper. Saute over low heat for 10 minutes or until vegetables are almost tender; blend in cornstarch. Drain syrup from pineapple; add water to make one cup; stir all liquids in pan until sauce thickens, boil 1 minute. Add ham and almonds. Heat through. Serve with seasoned cooked rice, and shredded coconut. Drizzle 1 Tbsp of honey.

Shepherd's Pie

1 lb ground beef)	
¼ cup green pepper)	Brown together
¼ cup onions)	

<table>
<tr><td>1 can tomato soup, undiluted)</td><td rowspan="4"></td></tr>
</table>

1 can tomato soup, undiluted)
¼ tsp salt   )
1 can green beans, undrained   ) Simmer for one half hour
1 Tbsp honey   )

Put in a casserole and spread mashed potatoes on top, and sprinkle with grated cheese.

Bake 15 minutes in a 425 degrees F. oven. Drizzle with 1 Tbsp honey.

Liver in Sweet and Sour Sauce

6 large, thin slices of liver	3 Tbsp oil
3 Tbsp whole wheat flour	1 Tbsp honey
1 egg, beaten	juice & rind of one lemon
4 Tbsp wheat germ	3 Tbsp parsley, minced

Dredge the liver in flour and dip in egg. Roll in wheat germ. Broil quickly on both sides in oven. (Keep the liver warm.) In sauce pan, blend oil, honey, lemon juice and rind. Heat thoroughly. Pour over liver. Garnish with parsley. Serves 6.

CHAPTER 7

CAKES

Carrot Cake

1½ cups salad oil	2 tsp cinnamon
4 eggs	½ tsp salt
½ cup honey	2 tsp soda
1½ cups raw sugar	5 cups finely grated carrots
3½ cups whole wheat flour	(more if using pulp from
½ tsp allspice	carrot juice)

Combine and mix thoroughly (about 1 minute on high speed with electric beaters): oil, eggs, sugar, honey. Sift in flour, spices and salt. Blend in grated carrot and bake at 350 degrees F. for 40 to 45 minutes.

Icing

4 oz. cream cheese	1 cup powdered non-instant
½ stick butter	milk
1 tsp vanilla	¼ cup honey
1 tsp maple flavoring	add halved walnuts and
1 cup powdered sugar	coconut

Mix all together. Spread on cooled cake (add a little milk if necessary to spread easier).

Honey Sponge Cake

1 cup honey	1/3 cup orange juice
½ cup brown sugar	½ tsp cloves
6 egg yolks	1 tsp vanilla
1¾ cups whole wheat flour	6 egg whites
sifted before measuring	1 tsp cream of tartar

Beat honey, sugar smooth. Add egg yolks one at a time and beat well until creamy and thick. Add orange juice, cloves and vanilla. Add flour and beat well. Beat egg whites and cream of tartar until very stiff (for at least 5 minutes). Fold into above mixture.

Bake in ungreased angel food pan at 325 degrees F. for 50 to 60 minutes or until cake springs back from touch. Remove from oven and turn upside down to cool 2 hours before removing from pan.

Fresh Apple Fruit Cake

1 cup boiling water	2 tsp cinnamon
1 pound raisins	½ tsp cloves
¾ cup shortening (part butter)	4 tsp baking powder
1¼ cups honey	½ tsp salt
3 eggs	3 cups grated raw apple
4 cups sifted flour	2 cups broken walnuts
1 tsp nutmeg	1 to 2 pounds fruit cake mix

Pour boiling water over raisins and allow to stand while thoroughly creaming together shortening, honey and eggs. Sift together dry ingredients. Drain raisins and spread on paper towel. Sprinkle a little of the flour mixture over raisins to keep them from sinking to bottom of cake. Add the raisins. Stir dry ingredients into batter alternately with grated apple. Fold in walnuts and fruit mix. Pour into two loaf pans that have been greased and floured. Bake in 300 degree F. oven for 60 min. or until done. Makes 2 large loaves.

Apple Pudding Cake

2 cups apples grated	¾ cup nuts
2 eggs	½ tsp salt
2 tsp vanilla	¼ cup oil
1½ cups whole wheat	¾ cup honey
2 tsp baking powder	

Blend honey, oil, beaten eggs, vanilla, add sifted flour, baking powder, salt, nuts, and apples. Bake about 30 minutes at 350 degrees F. Use with a sauce.

Pineapple Upside-Down Cake

Topping

2 Tbsp margarine	8 or 9 pineapple slices
1/3 cup honey	drained

Melt margarine and honey in bottom of 9" round or 8 x 8" square cake pan. Place pineapple halves around.

Batter

2 eggs
½ cup brown sugar
1 tsp vanilla
1¼ cup whole wheat or
  white flour

¼ tsp salt
1½ tsp baking powder
2 Tbsp shortening
½ cup warm milk

Mix eggs, sugar and vanilla well. Add shortening to warm milk—add to eggs and sugar alternately with flour, baking powder and salt. Pour batter over pineapple and bake at 350 degrees F. for 30 minutes.

Variations: Peaches, apricots or apples can be used.

CHAPTER 8

CANDIES

Honey Caramels

2 cups honey	3 Tbsp butter
¾ cup evaporated milk	pinch of salt
1 tsp vanilla	1 cup chopped nuts or more

Mix honey and evaporated milk together. Cook, stirring constantly, to a firm ball stage, at 255 degrees F. Stir in butter and nuts and salt. Pour onto a buttered square cake pan. Cool and cut into pieces.

Caramel Apples

Dip apples (dry and crisp) into cooked caramel (the above recipe) and spoon excess off bottom before drying on greased cookie sheets.

Basic Recipe for Balls

½ cup water	1 Tbsp butter
1 cup honey	½ tsp salt
1 cup brown sugar	1 tsp vanilla

Cook water, honey and brown sugar to 260 degrees F. or very firm ball stage. Remove from heat and add butter, vanilla, and salt.
Pour over 4 cups of any of the following:
1. popcorn; 2. puffed rice and peanuts; 3. puffed wheat and sliced dates; 4. puffed wheat and toasted almonds; 5. puffed millett and coconut.

Bits O Honey (Nutty Taffy)

1 cup honey	½ cup peanut butter
1 tsp vanilla	¾ cup roasted peanuts (unsalted)

Roast raw peanuts in 350 degree F. oven for 15 or 20 minutes. In sauce pan, cook honey (medium heat) to the medium ball stage. Remove from heat, add vanilla. Pour onto a buttered platter. Let cool just enough to work with hands. Add peanut butter. Butter hands and pull until lighter in color. Work in peanuts and press or roll out on buttered surface until ¼-inch thick. Cut in pieces and wrap in waxed paper. Makes about 24 pieces.

Honey Taffy

½ cup honey	2 Tbsp margarine
¼ cup corn syrup	1/8 tsp salt
½ cup water	1½ cup sugar

Cook to 270 degrees F., add ¼ tsp peppermint flavoring. Cool on buttered platter. Stretch.

Honey Popcorn Balls

1 cup honey	½ cup water
1 cup brown sugar	½ cup canned milk
¾ cup white sugar	1 tsp vanilla

Cook to soft ball stage, add vanilla, then pour over popped corn. Let set few minutes before forming balls. (Good for mailing).

Missionary Candy (Crisp)

1¼ cup oatmeal	½ cup margarine or butter
1 cup flaked coconut	½ cup honey
1 cup walnuts chopped	½ cup brown sugar
½ cup toasted wheat germ	1/3 cup sesame seeds
½ cup snipped dried apricots	1 tsp cinnamon

Combine margarine and honey and sugar in pan, heat. Stir into rest of ingredients. Spread in a 13 x 9" pan; bake at 350 degrees F. for about 25 minutes, stirring two or three times; turn onto greased foil; break in pieces. Good for mailing.

Nut-Honey Treat (Crisp)

2 cups honey	1 cup sunflower seeds
1½ cup sesame seeds	1 cup coconut
1½ cup wheat germ	1½ cup peanuts

Simmer above ingredients for 1 hour, or until hard ball stage. Pour into buttered pan; cool, and cut into squares.

Creamed Taffy Candy

1 cup brown sugar	1 cup cream
2 cups honey	

Mix all ingredients and cook slowly to the hard ball stage, at 250 degrees F. Pour onto buttered platter. Cool, butter hands and pull to the golden color. Cut in bite size pieces.

Andersen Crunch

1 cup honey	2 cups nuts, almonds,
1 cup molasses, mild	peanuts or walnuts
1½ cup brown sugar	3 Tbsp margarine or butter
1½ cup sesame seeds	1½ cup soy powder
1½ cups sunflower seeds	2 tsp vanilla

Cook molasses, honey, brown sugar to 250 degrees F., then add other ingredients. Mold into balls or pour into flat pan and cut in squares.

Uncooked Honey Peanut-Buttered Candy

1 cup honey	1 cup peanut-butter
1 to 1½ cups powdered milk	
(non-instant)	

Mix all ingredients together thoroughly, (slightly warm makes it easier). Pour into a square pan one-half inch thick, and cut into squares. Plain candy rolled into balls and pressed flat can be trimmed with nuts like sunflower seeds in a flower-like design or any original design, or mix with other nuts.

Mold into balls dipped in coconut, or roll in coconut. Mix a portion in Rice Krispies, rolled oats, rolled wheat, or granola. A dried date pressed into a ball, or any other dried fruit. Figs, etc. Carob chips are delicious. Walnut meats are excellent. Make sesame seeds in a nest. Cover a Brazil nut with the candy. Roll a ball in popped wheat. Good with a bite-sized piece of candied pineapple embedded in a ball. Also good as a sandwich of dried banana between layers of candy and vice versa.

Super Fudge

1 cup honey	1 cup dry milk—optional
1 cup peanut butter	1 cup sunflower seeds—
1 cup wheat germ—optional	optional
1 cup carob powder—optional	½ cup coconut—optional
1 cup soyflour—optional	Fruit, dates or nuts can
1 cup sesame seed—optional	be added

Mix honey and peanut butter. Add rest of ingredients. Pour into a square pan, cut into squares and refrigerate to harden. Keep refrigerated.

Honey Marshmallows (Uncooked)

¾ cup raw or brown sugar	¼ tsp salt
1 cup honey	1 tsp vanilla
1½ Tbsp unflavored gelatin	2 cups nuts, walnuts or
1/3 cup water	pecans

Pour water in small saucepan and sprinkle gelatin on top. Let stand 5 minutes. Place pan over medium heat and stir only until dissolved. Add raw sugar and continue until it is dissolved. Do not boil. Add honey, salt, and vanilla. Transfer all to a bowl and beat with electric beater until cool and tacky. Add nuts, fruit etc. Any of the following are good: coconut, carob chips, chopped dried apricots, raisins or currants. Sprinkle powdered milk or powdered sugar on bottom of square cake pan. Pour candy into pan and spread evenly. Let stand a few hours before cutting with a wet knife.

Roll squares in toasted coconut or popped wheat, or dip squares in carob coating. Roll dipped candy in toasted or plain coconut or best yet, toasted, ground almonds. Have fun!

All-Day Suckers

1 cup honey	1½ cups dry milk

Cook honey to hard crack stage at 285 degrees F. Remove from heat, and immediately add enough dry milk to make a stiff ball. Smooth out the lumps. Roll into small balls, insert sucker sticks and allow to harden.

Try this: Before shaping into balls, add vanilla or other flavoring or coloring.

Basic Candy (Use Basic Recipe, p. 43.)

1½ cup non-instant dry milk   ½ cup warm honey

Stir and knead enough dry milk into the honey to make a very firm ball. This can be rolled paper-thin, molded and shaped for crafts. For variety, try adding peanut butter, substitute molasses for the honey; add nuts, dried fruit, flavorings and coloring.

Vary the shape. Roll and cut into pieces the size of pine nuts, small mints, or small oblongs. Or grind pieces of the firm candy in the meat grinder, and as they come through form them in pointed chocolate-drop shape and size or in small circles.

Variations: (1) Substitute molasses for honey in basic recipe, p. 43. (2) Color candy red or green and use raspberry or mint flavoring. (3) Add peanut butter to basic recipe. (4) Use equal parts of soybean flour, dry milk and peanut butter, moisten with lemon juice and honey and roll out. Mold in pecan-roll shape and cut in slices with a string. (5) For macaroons, add fine coconut to basic recipe. Shape in small balls, flatten with thumb and forefinger and brown under broiler.

Carob Fruit Candy

2 cups noninstant milk   1 tsp vanilla
½ cup carob powder   1 cup raisins, dates or
2 Tbsp butter   other dried fruits
1 cup honey   1 cup nuts
½ cup coconut

Mix milk and carob powder together, then the other ingredients. Mold into balls and roll into the coconut. It could be made into a roll and cut in slices or in a square pan and cut in bite size squares.

CHAPTER 9

COOKIES, FRUIT BARS, AND DOUGHNUTS

Honey Nuggets

1/3 cup shortening	½ tsp baking soda
½ cup honey	½ tsp salt
1 egg	1 package (6 oz.) semi-
½ tsp vanilla extract	sweet chocolate chips
1¼ cups sifted flour	½ cup chopped nuts

Cream together shortening and honey. Add egg and vanilla. Beat until light and fluffy. Sift together flour, soda and salt. Add to first mixture along with chocolate chips and nuts. Drop by the teaspoonful onto greased cookie sheet. Bake in a moderate oven (375 degrees F.) 10 to 12 minutes. Makes 4 dozen cookies.

Honey Valentine Cookies

½ cup shortening	2½ cups white flour
¾ cup honey	1 cup whole wheat flour
1 egg	½ tsp soda
¼ cup milk	4 tsp baking powder
1 tsp vanilla	½ tsp salt

Cream shortening and honey together. Beat the egg, and cream again. Add milk and vanilla. Combine the dry ingredients, sift, and stir into the first mixture. Roll the dough to about ¼ inch thickness and cut with a heart-shaped cookie cutter. Place on greased tin, and bake in a 350 degree F. oven. Makes about 36 cookies.

Honey Lace Cookies

½ cup shortening	½ tsp baking powder
½ cup sugar	¼ tsp salt
½ cup liquid honey	1 cup quick-cooking rolled
1 egg	oats
1 cup sifted all-purpose flour	1 cup shredded coconut
½ tsp soda	½ cup chopped nuts

Cream shortening, sugar and honey together until light. Add egg and beat well. Sift flour, soda, baking powder, and salt together. Combine with creamed mixture. Stir in the rolled oats, coconut and nuts. Drop by tablespoons on a greased cookie sheet 2 inches apart. Bake at 350 degrees F. for 15 minutes, or until cookies are golden brown. Remove from pan while warm. Makes 30 cookies.

COOKIES AND BARS

Best Honey Cookies

5 cups whole wheat flour	1 cup shortening
1 tsp salt	2 eggs
1 tsp baking powder and	2 cups honey
½ tsp soda	1 tsp vanilla
1 tsp each cinnamon, ginger	1 cup milk
½ tsp each cloves, nutmeg	

Sift dry ingredients together until light and fluffy. Combine eggs, honey, shortening and vanilla. Beat well. Add milk alternately with dry ingredients. Chill dough. Roll out and cut with glass or form in balls and flatten slightly. Bake 8 to 10 minutes at 375 degrees F. if rolled, or 15 minutes if in balls.

Variations: Add raisins or dates or coconut, or while warm roll in powdered sugar and cinnamon, or before baking form into balls and roll in fine coconut.

Honey Apple Cookies

¾ cup honey	1½ tsp cinnamon
½ cup shortening	½ tsp allspice
½ tsp vanilla	1½ cups finely chopped
2 eggs	apple
2 cups sifted whole wheat	1 cup chopped raisins and
flour	nuts mixed
1 tsp soda	

Cream shortening and sugar. Add eggs and vanilla and beat well. Add apples and mix well. Sift dry ingredients together and add, beating well by hand. Add raisins and nuts last. Drop by teaspoonfuls onto a greased cookie sheet. Bake at 350 degrees F. for about 15 minutes.

Carrot Bars

1 cup water	1 tsp each cinnamon,
1 cup raisins	nutmeg, and soda
¾ cup honey	2 cups whole wheat flour
1/3 cup shortening	½ tsp salt
1 large grated carrot	
(1½ cups, packed)	

Cook raisins and water until raisins are soft. Cream together honey and shortening. Add grated carrot and stir. Then add sifted dry ingredients along with raisins and water mixture. Add nuts if desired. Spread into greased 9-inch by 12-inch pan. Bake at 350 degrees F. for 25 to 30 minutes. Frost with 1 cup powdered sugar, ½ cup regular dry milk, ½ tsp vanilla and milk to thin consistency. Makes 2 dozen 2 inch by 2 inch bars.

Honey-Covered Raisins

½ cup honey	1 cup toasted fine coconut
4 cups raisins	
(bleached white raisins if available)	

Warm honey and stir raisins into it. Then lift raisins out with slotted spoon, drop into coconut. Mix until well coated. Spread out on wax paper. Separate and cool. Do the same with figs, prunes, dates or dried cherries.

Fruit Squares

1 cup raisins	1 cup dates
1 cup dried prunes	1 cup nuts
1 cup figs	1 cup hard granulated honey

Grind all fruit. Chop nuts. Blend both with hard honey. Press into square pan. Cut in squares and coat with coconut or powdered sugar.

Fruit Crisp

3 cups canned fruit	1 cup flour
(pears are especially good)	5 Tbsp butter
4 cups fresh fruit	2 Tbsp raw sugar
1 cup gluten crunch	

Place fruit which has been cut into bite-size pieces on bottom of loaf pan. (If it is canned fruit, drain well.) Mix remaining ingredients and sprinkle over fruit as evenly as possible. Bake 40 minutes at 350 degrees F. Serve with whipped cream or milk. (Good also for a breakfast dish.)

Hint: For a delicious apple crisp, lightly sprinkle cinnamon sugar (half & half mix) over peeled and cut apples before putting on the topping.

Honey Apple Crisp

2 cups pared and sliced apples
2 Tbsp sugar
1½ tsp lemon juice
¼ cup liquid honey

¼ cup all-purpose flour
2 Tbsp brown sugar
1/8 tsp salt
2 Tbsp butter or margarine

Place apples in a shallow baking dish. Combine the sugar, lemon juice, and honey. Spread over apples. Mix the flour, brown sugar and salt; cut or work in the butter or margarine until mixture is crumbly. Cover apples with the flour mixture and bake at 350 degrees F. for 30 to 40 minutes, or until the apples are tender and the crust is brown. This is good served with whipped cream and a dash of cinnamon on top. Makes 4 servings of about ½ cup each.

Honey Cakes

1 egg
¾ cup brown sugar
½ cup honey
½ cup dark molasses
3 cups sifted flour
1¼ tsp cinnamon
1¼ tsp nutmeg

½ tsp cloves
½ tsp allspice
½ tsp soda
½ cup chopped mixed candied fruits and peels
½ cup slivered blanched almonds

Beat the egg, add brown sugar and beat mixture until fluffy. Stir in the honey and molasses. Sift together the dry ingredients and add to first mixture. Mix well. Stir in fruits, peels and nuts. Chill several hours, or overnight. Roll ¼ inch thick on floured surface. Cut into rectangles 3½ x 2 inches. Bake on greased cookie sheet at 350 degrees F. for 12 minutes. Cool slightly before removing from pan. Makes about 2 dozen.

Carob Coconut Cookies

¾ cup honey
½ cup oil
¼ cup milk
2 eggs
3 Tbsp carob powder
1½ cups oatmeal
1 cup whole wheat flour

2 tsp baking powder
½ tsp salt
¼ tsp nutmeg
1 tsp cinnamon
1 tsp vanilla
½ tsp almond

Mix together honey, oil, milk, eggs, flavorings. Add sifted dry ingredients. Drop with spoon, bake at 350 degrees F. 10 or 15 minutes. May need more flour to hold shape.

Filled Cookies

Filling:

1 cup water
2 cups dates, raisins or any softened, chopped dried fruit

½ cup honey, or ¾ cup raw sugar
1 cup nuts, chopped

Mix water, dried fruit and honey. Cook for 5 minutes, stirring continually. Add the nuts and allow the mixture to cool.

Dough:

1 cup vegetable oil
1½ cups honey
2 eggs
½ cup water
1 tsp vanilla

4 cups whole wheat flour
½ tsp salt
1 tsp soda
1 tsp cinnamon

Cream shortening, honey and eggs together. Add water and vanilla. Sift the dry ingredients together, and add to the creamed mixture. Place 1 teaspoon dough on cookie sheet. Add 1 teaspoon filling on top of this; then another teaspoon of dough to cover the filling. Place 2 inches apart. Bake 10 minutes at 375 degrees F.

Glazed Honey Doughnuts

2/3 cup milk
2 Tbsp shortening

3 cups sifted all-purpose flour

¼ tsp salt
1/3 cup liquid honey
½ cake compressed yeast
   crumbled, or ½ package
   active dry yeast

1 egg, beaten
¾ tsp cinnamon
¼ tsp nutmeg

Scald milk, add shortening, salt, and 1 tablespoon of honey. Cool to lukewarm. Stir in the yeast, add 1 cup flour, and beat well. Set this sponge in a warm place for 1 hour (or until mixture is full of bubbles). Combine remaining honey with the egg and spices, then stir into the sponge. Add remaining flour. Turn out onto floured board and knead 1 minute. Place dough in bowl, cover and let rise until double in bulk (about 1½ hours). Turn dough out on floured board, and roll ½ inch thick. Cut with floured doughnut cutter and let rise until light (about 1¼ hours). Drop with raised (top) side down into deep fat and fry at 360 degrees F. for 1 minute on each side. Drain, then cover with a honey glaze. Makes 16 doughnuts.

Honey Glaze

1/8 tsp unflavored gelatin
½ cup water
1 Tbsp liquid honey

2¼ cups confectioner's sugar
1/8 tsp salt
¾ tsp vanilla

Combine gelatin and water, add honey, and heat over hot water until warm. Add sugar, salt and vanilla; stir until smooth. Keep the glaze warm over hot water, dip the doughnuts in it, then place them on a rack to dry. Makes enough for 16 doughnuts.

CHAPTER 10

PIES

Pumpkin Pie

1 No. 2½ can of pumpkin	4 eggs
1 cup honey	1 tsp each salt, cinnamon,
2 Tbsp flour	nutmeg
2 cups milk	¼ tsp ginger
1 cup canned milk (condensed)	

Mix honey, pumpkin, seasoning and flour together with milk and condensed milk. Heat to boiling point and remove from stove. Add the eggs which have been beaten. Pour mixture into pie crusts which have been allowed to bake 4 minutes. Return to oven and bake at 350 degrees F. until a knife inserted comes out clean. This requires about 45 minutes.

Lemon-Honey Chiffon Pie

1 Tbsp unflavored gelatin	½ cup lemon juice
¼ cup cold water	1 tsp grated lemon rind
4 eggs separated	9-inch baked pastry shell
¾ cup honey	1 cup heavy cream, if
½ tsp salt	desired

Soften gelatin in the cold water; set aside. Beat egg yolks and combine with honey, salt, and lemon juice and rind. Cook over hot water until thick, stirring constantly. Add gelatin and stir to dissolve. Remove from the heat and cool. Beat egg whites until stiff, then fold into the custard mixture. Turn into a 9-inch baked pastry shell. Chill until firm. Top with whipped cream before serving, if desired. Whipped cream may be sweetened with honey.

Pumpkin Pecan Pie

3 slightly beaten eggs	½ cup Karo syrup
1 cup pumpkin	½ tsp cinnamon
1 cup brown sugar, or	¼ tsp salt
1 cup honey	1 cup pecans (chopped)

Combine first 7 ingredients. Pour into unbaked pastry shell. Top with nuts. Bake at 350 degrees F. for 40 minutes.

Pie Crust

2 cups whole wheat flour
1 tsp salt

5 Tbsp cold water
2/3 cup lard, cold

Sift flour and salt. Remove 1/3 of this flour and make a paste by adding 5 tablespoons water. Blend lard with remaining flour until size of peas. Add paste, and stir to a ball. Roll out and chill in pans. Bake at 425 degrees F. for 15 minutes, and then turn down to 350 degrees F. for another 30 minutes. This recipe makes 3 crusts.

Rhubarb Pie

3 cups rhubarb,
   diced
1 cup water
3/4 cup honey

2 Tbsp cornstarch
1 tsp vanilla
2 Tbsp butter

Line a nine inch pie plate with pie crust and moisten edges with water. Fill with rhubarb. Combine water, honey, cornstarch, vanilla, and butter and cook until clear. Cool slightly and pour over rhubarb. Fit on a top crust, press edges with a fork and trim off crust. Slash to allow steam to escape. Bake 35 to 40 minutes in hot oven, 400 degrees F. for first 10 minutes, then 350 degrees F. for remaining time.

Apple-Cherry Pie

3 large cooking apples
6 Tbsp butter
2½ cups pitted sour red pie
   cherries, fresh or canned
3 Tbsp tapioca

¾ cup honey
2 Tbsp flour
2 tsp cinnamon
½ tsp nutmeg

Make pastry for two-crust pie. Melt 2 tablespoons butter and brush on bottom of pastry shell. Mix all the ingredients and put into shell. Top with dots of remaining butter. After top crust is added to pie, rub crust with cream or evaporated milk, and sprinkle with mixture of ½ teaspoon sugar and ¼ teaspoon cinnamon. Bake in 425 degree F. oven for 15 minutes, then turn oven to 300 degrees F. and bake for 40 minutes.

CHAPTER 11

PUDDINGS AND OTHER DESSERTS

Danish Sweet Soup

2 cups dried prunes, soaked	½ cup minute tapioca
3 cups raisins	2 tsp vanilla
1 cup dried apples	4 Tbsp butter
¾ cup honey	2 sticks cinnamon or
1/8 tsp salt	2 tsp ground cinnamon

Cover fruit with water and boil until soft. Add honey, tapioca, butter, vanilla and cinnamon. Add enough more water to make 1½ quarts of liquid. Serve hot or cold. Use any dried fruit combination. Apricots and peaches are a good color combination.

Red Mush

2 cups red currants	3 cups water
(any berry can be used)	1/3 cup minute tapioca
½ cup honey	

Crush berries, add water and honey. Bring to a boil and thicken with tapioca.

Raspberry Rhubarb

4 cups cut rhubarb	½ tsp raspberry flavoring
1 cup honey	¼ tsp red coloring
2-3 cups water	

Combine all ingredients and bring to a boil. Reduce heat and simmer for 10 minutes. Cover pan and do not stir.

Variation: Omit flavoring and add 1 package raspberry jello at end of cooking.

Honey Fruit Cobbler

1 egg, separated	¼ tsp salt
1 cup honey	2 tsp baking powder

¼ cup margarine
½ cup milk
½ tsp vanilla
1 cup flour

3 cups berries, cherries, apples, etc.
1 cup fruit juice

Blend ½ cup honey with the margarine. Add egg yolk, milk and vanilla. Add dry ingredients and beat well. Add beaten egg white last. Drizzle remaining half cup honey over fruit. Spread batter in cake pan first. Place fruit on top of batter. Pour 1 cup juice over all. Bake at 350 degrees F. for 30 minutes.

Millet Pudding

3 cups juice (apricot is good)
1 cup millet

2 cups pineapple
4 Tbsp honey

Mix juice and millet in double boiler. Cook until millet is soft (about 1½ to 2 hours). Add crushed pineapple.

Variations: Can use other kinds of juices. Add nuts, raisins, dates, coconut, etc. Top with whipped cream. When cool, add 4 or 5 bananas.

Honey-Peanut Butter Custard

1-1/3 cups skim milk
1/3 cup peanut butter
2 eggs

3 Tbsp honey
¾ tsp salt

Add the milk gradually to the peanut butter, stirring until smooth. Beat eggs slightly; add honey and salt, and stir to mix. Combine the two mixtures. Pour into a small baking dish. Set baking dish in a pan of hot water and bake at 325 degrees F. about 30 minutes, or until custard is set. Serves 4.

Honey Delight

1 pkg. lemon or orange gelatin
½ cup boiling water
½ cup honey

Juice of ½ lemon
1 can evaporated milk, chilled
Graham crackers

Dissolve gelatin in boiling water. Add honey and lemon juice and mix well. Fold in evaporated milk that has been chilled and whipped. Pour into pan that has been lined with crushed Graham crackers. Sprinkle cracker crumbs on top, refrigerate a few hours.

Yummy Rice Pudding

¾ cup honey
Pinch salt
3 cups milk
3 slightly-beaten eggs

1½ tsp vanilla
3 cups cooked brown rice
1 tsp grated lemon rind
1 cup raisins

Stir well and pour into greased baking dish, set in flat pan containing an inch of hot water. Bake for one hour at 325 degrees F. or until set.

Apple Ring

2 cups flour
2 tsp baking powder
2 Tbsp oleo

Enough milk to make dough
   for rolling, ¼-inch thick
Grated apples to cover dough

Sprinkle with cinnamon and roll up like jelly roll and slice. Put in baking dish and pour syrup over. Bake 20 minutes at 350 degrees F.

Syrup

1 cup honey
1 cup brown sugar

2 cups water
2 Tbsp oleo

Mix and bring to a boil.

Lemon-Honey Crisp

6 Tbsp oleo
½ cup honey
1 cup flour
½ tsp soda

½ tsp salt
½ cup coconut
¾ cup crushed soda
   crackers

Combine and press in pan. Bake 8 minutes at 350 degrees F.

1 cup hot water
1/3 cup honey
¼ cup brown sugar
3 Tbsp cornstarch

¼ tsp salt
2 yolks, beaten
½ tsp lemon peel, grated
½ cup lemon juice

Mix sugar, cornstarch, and salt in a pan; add water slowly. Cook until it boils, remove and slowly add yolks, lemon juice and peel. Pour over crust and top with crumbs. Bake at 325 degrees F. for 25 minutes. Serve with whipped cream.

Carob Pudding

1 cup water	3 Tbsp carob powder
3 eggs	1 tsp vanilla
¼ cup oil	¼ cup cream or canned milk
¼ cup honey	

Bring water to boil; stir in beaten eggs and stir until as thick as a thin mush. Pour into blender, add other ingredients. Vary to suit taste. May be frozen for fudgsicles, used as a pie filling, or pudding, or topping for ice cream. Delicious with whipped cream when served warm.

Honey Rice Pudding

2 cups cooked rice	3 eggs, slightly beaten
3 cups milk	1 cup raisins
2/3 cup honey	nutmeg

Mix rice, milk, and honey. Add the eggs. Stir in the raisins. Sprinkle nutmeg on top. Bake in greased baking dish for about one hour.

Poor Man's Pudding

¼ cup shortening	½ tsp salt
1½ cups whole wheat flour	½ cup milk
½ cup honey	½ cup raisins
½ tsp soda	½ cup nuts (optional)
1 tsp baking powder	½ tsp cinnamon

Blend shortening and honey; add milk, then dry ingredients and raisins. Spread out in cake pan. Dissolve 1 cup brown sugar in 2 cups hot water. Pour over batter and bake at 350 degrees F. for 30 minutes.

Variation: Dates may be substituted for the raisins.

Honey Bread Pudding

2 cups day-old bread cubes	2 eggs, beaten
¼ cup honey	½ tsp vanilla
2 Tbsp margarine	1-2/3 cup hot milk
1/8 tsp salt	½ cup raisins (optional)

Place bread cubes in small baking dish. Combine the honey, margarine, salt, eggs, and vanilla. Slowly stir in the milk. Pour the mixture over the bread. Set baking dish in a pan of hot water and bake at 350 degrees F. for 30 to 40 minutes or until pudding is set.

Baked Apples

Core apples to a depth allowing the bottom to be closed. Place in a baking dish with sufficient water to sustain moisture. Bake until tender, drizzle honey over apples and return to oven for 10 minutes. You will be delighted with the aroma and delicacy of the honey as if it were fresh from the honey comb. Add a dash of cinnamon if desired.

A Quick Icing or Topping for Apple

Beat 2 egg whites until stiff. Add 1 tsp honey and ¼ cup whipped cream. Beat all stiff.

Variation: 2 oz. Philadelphia Cream Cheese, 1 tsp honey, and ½ tsp lemon juice. Blend all well.

Another Variation: (for breakfast) Mix the following together for a topping:

2 Tbsp raw or brown sugar	½ tsp cinnamon
2 Tbsp honey	2 Tbsp melted butter
½ tsp nutmeg	½ cup Granola

Honey Apple Betty

2 Tbsp margarine or butter	¼ cup liquid honey
1 cup soft bread crumbs	¼ cup warm water
1 Tbsp lemon juice	¼ tsp cinnamon
½ tsp grated lemon rind	¼ tsp nutmeg
2 cups pared and sliced apples	

Serves four.

Melt butter or margarine and stir into bread crumbs. Add lemon juice and rind to apples. (Omit lemon if apples are tart.) Mix honey and water. Place a layer of crumbs in a greased baking dish and cover with a layer of apples. Moisten with honey mixture and sprinkle with part of the seasonings. Repeat layers, ending with bread crumbs as the top layer. Bake at 350 degrees F. for 30 to 45 minutes until crumbs are well browned. Cover baking dish for the first 15 minutes. Makes 4 servings of about ½ cup each.

Honey Apple Crisp

4 cups pared, sliced apples	4 Tbsp brown sugar
3 tsp lemon juice	¼ tsp salt
½ cup honey	4 Tbsp butter or margarine
½ cup flour	

Place apples in a shallow baking dish. Combine the lemon juice and honey. Spread over apples. Mix the flour, brown sugar, and salt. Cut or work in the butter or margarine until mixture is crumbly. Cover apples with the flour mixture and bake at 350 degrees F. for 30-40 minutes or until the apples are tender and the crust is brown. Good served with whipped cream and with a dash of cinnamon on top. Serves 8.

CHAPTER 12

ICE CREAM

Honey and Whey Ice Cream

1 cup milk	3 Tbsp to ¼ cup honey
2 eggs	1 cup whipping cream
¼ cup whey	1 Tbsp vanilla

Scald 1 cup milk. Blend in blender 2 eggs, ¼ cup whey, 3 Tbsp to ¼ cup honey according to taste. Add scalded milk to mixture, stir vigorously by hand, return to low heat and cook stirring 5 min. or until mixture coats the spoon. Cool, pour into chilled tray to mushy stage. Chill a bowl and beaters. Remove custard from freezer. Whip until smooth in chilled bowl. Whip one cup light whipping cream with 1 Tbsp vanilla. Fold custard into whipped cream. Freeze until firm.

Note: to keep a smooth texture and to prevent the ice cream from becoming grainy, be sure that (1) both mixtures are the same temperature, (2) don't over-freeze, and (3) use up fairly soon.

Fresh Strawberry Ice Cream
(Summer Freezer Special)

4 eggs	1 tsp salt
2¼ cups honey	1 Tbsp vanilla
4 cups milk	red food coloring, if desired
2 cups heavy cream (pint)	2 cups crushed strawberries
2 cans (13 oz.) undiluted milk	

In a large bowl, beat eggs slightly, gradually add honey, and mix well. Add milk, cream, evaporated milk, salt, flavoring and coloring. Pour into freezer can. Assemble freezer and pack with ice and salt. When partly frozen, add crushed strawberries; continue freezing. Remove dasher and pack for ripening for 2 or 3 hours. For firmer texture reduce the amount of honey and add sugar. Makes 1 gallon.

Vanilla Ice Cream

½ cup raw sugar
¼ cup honey
4 egg yolks
2 tsp vanilla

dash of salt
1½ cups heavy cream,
   whipped
1 Tbsp cornstarch

Mix sugar, honey, and cornstarch. Boil for 3 minutes, stirring continually. Put in blender and add eggs, salt and vanilla. Blend well. Fold this mixture into the whipped cream. Pour into refrigerator tray. Cover with waxed paper. Freeze 2-3 hours.

Raspberry or Strawberry Ice Cream

1 Tbsp lemon juice
3 Tbsp protein powder
2/3 cup sweetened condensed
   milk

1-10 oz. pkg. frozen fruit
1 cup heavy cream, whipped

Put all ingredients into blender except the milk and cream. Mix for 20 seconds. Slowly add milk. Fold this mixture into the whipped cream. Freeze.

Carob Ice Cream
(Freezer)

¾ cup evaporated milk
1 cup reconstituted
   powdered milk
1 tsp carob, powdered

4 Tbsp melted dipping carob
3 Tbsp honey
¼ tsp maple flavoring

Put all ingredients in blender and mix at high speed for 10 seconds. Mix 30 seconds or until smooth. Pour into container and freeze. This can be whipped again before serving.
   Hint: Add mint flavoring.

Ice Cream Tiger

1 Tbsp Tiger's Milk
  protein powder
1 cup starch water
2 cups frozen fruit

1 tsp honey
1 tsp vanilla
1 rennet tablet

(This is more like a sherbet.) See Rennet "Junket" instructions for freezing.

Hint: Sprinkle any of the above ice creams with 1 cup coconut, sweetened or unsweetened, 3 Tbsp melted butter or margarine, and 4 Tbsp honey or raw sugar.

Carob Ice Cream
(Refrigerator)

1 quart milk
1 cup cream or
  evaporated milk
¼ cup carob powder or
  to suit taste

½ cup honey
2 tsp vanilla
½ cup soy powder (optional)

Blend and freeze in ice trays. Put thru blender. Thaw and whip with electric beater.

Angel Yule Log
(Holiday Eye Catcher)

4 egg yolks
¾ cup honey
¼ tsp salt
½ cup lemon juice
1 tsp grated lemon peel
1 Tbsp plain gelatin

1/3 cup sweet sherry wine
4 egg whites
¼ cup honey
½ cup diced candied cherries
1 (12 oz.) baked angel food
  cake

Beat egg yolks well; continue beating while adding honey in a fine stream. Add salt, lemon juice and peel; cook and stir over low heat until mixture thickens. Soften gelatin in sherry; dissolve in hot mixture. Beat egg whites to stiff peaks. Continue beating while slowly adding ¼ cup honey. Gently fold into custard; add cherries.

Break angel food cake into bite-size pieces. Combine with custard. Turn into a buttered loaf pan about 10 x 4 x 3 inches

(1¾ quart capacity), or use an angel food pan.

Chill in refrigerator 6 hours or overnight. Unmold and garnish if desired with additional whipped cream and cherries. Makes 12 servings.

CHAPTER 13

JAMS AND JELLIES

Pineapple Apricot Jam

3 cups pineapple, crushed & drained

2 cups ground apricots—fresh or dried

4 cups mild honey

pineapple liquid plus juice of 1 lemon and water to make up to 1 cup

1 box pectin powder—1¾ oz.

Combine fruit, juice and pectin. Bring to a boil, stirring constantly. Stir in honey and boil 1 minute hard (2 minutes if stiffer jam is desired). Remove and stir for 5 minutes. Pour in sterile jars and seal.

Frozen Strawberry Jam

3½ cups mashed strawberries

¼ cup lemon juice

1 pkg. pectin powder 1¾ oz.

3½ cups mild honey

Combine mashed strawberries, lemon juice and pectin and let stand 30 minutes, stirring occasionally.

Add 3½ cups honey and warm to 100 degrees F., stirring to blend honey and fruit. Pour into sterilizied jelly glasses and freeze.

If used within 2 to 3 weeks it may be stored in refrigerator.

Fruit Juice Jelly

½ pint fruit juice

honey to taste

2 cups of water and agar-agar simmer for 15 min. until clear

Stir well, add honey to hot liquid, then juice and water. Pour into wetted individual molds and leave in a cool place to set. This jelly sets much more quickly than jelly made with animal gelatine and will be set as soon as cold.

Apricot Sauce, Jam or Jelly

4 cups ripe apricots

½ cup honey

Mash fruit and simmer 15 minutes. Seal in sterilized fruit jar, or cook in hot water bath 20 minutes.

Variation: Instead of apricots, try pears, peaches, or plums.

In canning your fruit add 2 Tbsp of honey instead of sugar. Your fruit will taste fresher.

Place fruit into sterilized canning jar. Add two or three Tbsp of honey (according to the sweetness of the fruit). Pour hot water over the fruit to fill jar within ½ inch. Seal and follow canning instructions.

Grape Jelly (Uncooked and Frozen)

5 cups grape juice (unsweetened)	2 pkgs pectin
1 cup water	2 cups honey
½ cup corn syrup	½ cup lemon juice
	4½ cups sugar

Combine juice, water, lemon juice. Add pectin slowly while stirring vigorously. Set aside 30 minutes. Stir occasionally.

Add corn syrup, stir. Add honey and sugar. Warm to 100 degrees F. Put in jars with lid. Freeze or use.

CHAPTER 14

OTHER HONEY RECIPES

Honey Fruit Leather

Select any ripe fruit, apples, peaches, pears, apricots, bananas etc. Blend well in a blender or a Champion Juicer. Add one Tbsp of honey and one Tbsp of lemon juice to a quart, depending on the sugar content of the fruit. Pour to ¼ inch thick on a medium-heavy plastic on a very smooth surface. Leave in hot sun to dry (one day will be sufficient) or dry in a dehydrator or oven (at low temperature, with the door ajar). Do not dry to a crisp; a small amount of fruit moisture will not cause mold.

Variations: A combination of fruits will make a delicious fruit leather.

Dipped Fruit Pieces

This is a recipe for honey syrup on dried fruits. Mix two parts water, one part honey or more, one tsp of ascorbic acid (or Fruit Fresh or a Vitamin C tablet). Dip fruit into syrup and place on dehydrator tray. When like soft leather, place in a sterilized bottle, not quit full. Seal and place in 150 degree F. oven for 20 to 30 minutes.

Variations: Especially good with apples (¼ inch slices). Empty one package Jello into 2 cups water, (orange flavor is especially good). Add honey to taste or, for a special treat, sprinkle dry Jello on dipped fruit before placing in the dehydrator. It makes a most delicious fruit leather. Use various flavors of Jello.

Granola

7 cups rolled oats	1 tsp salt
1 cup wheat germ	1 tsp vanilla
1 cup honey	1 cup oil
2 cups coconut	grated rind of 1 lemon
1 cup sesame seeds	1 cup soya flour
¾ cup nuts, chopped	¼ cup water

Bake in slow oven 200 degrees F. for 2 hours, stirring often. Add raisins or other dried fruits after baking. The following ingredients could be added.

Variations:

6 cups wheat flakes	2 cups yeast flakes
1 tsp maple	1 cup sunflower seeds
1 cup date granules	

Can be eaten as a snack dry or with milk (half and half) or cream.

Honey Buds

Honey buds is a product made from 96% pure honey and 4% whey from which moisture has been extracted, reducing it to small white flakes and pouring like sugar.

Artificial Honey

10 cups sugar (5 lbs)	2 cups water
¾ tsp alum	20 large rose petals
48 large or 60 small clover blossoms	

Any natural sweet or any fruits or equivalent. Mix ingredients. Boil gently 10 minutes. Let set for 15 minutes then strain.

Ten pounds of sugar makes a double batch, so double ingredients and let set another 10 minutes and you get 10 lbs of honey.

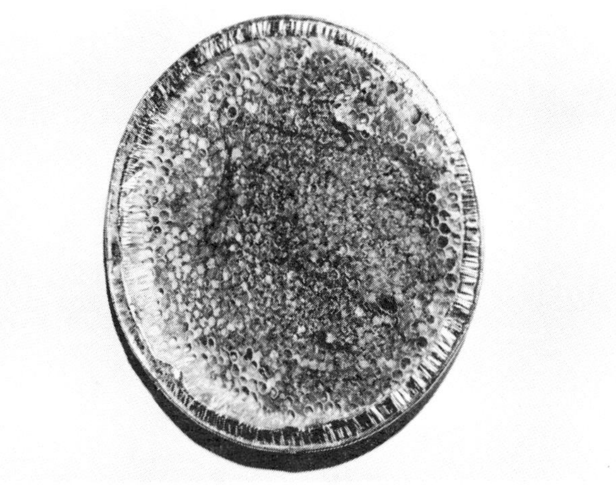

HONEY PIE

A unique way of serving comb honey to friends. Produced by placing two tin foil plates back to back in the hive. The two plates fasten into a standard frame to take the place of a regular foundation frame. Originated and presented to the author by Byron C. Peacock.

PART II

BACKYARD BEEKEEPING

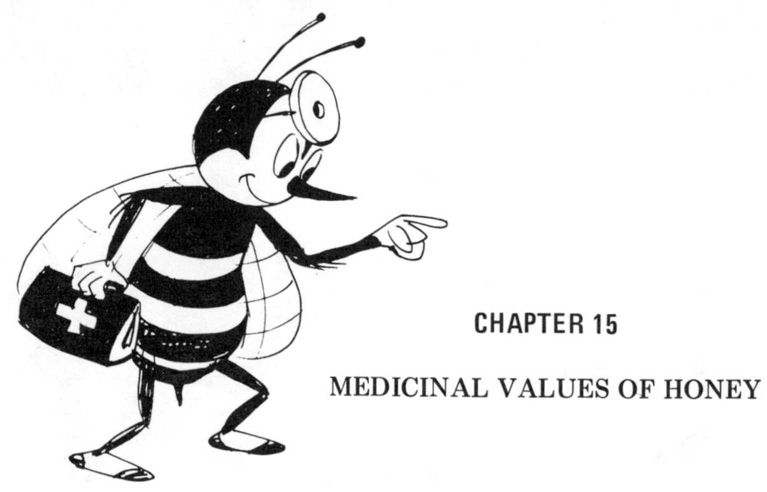

CHAPTER 15

MEDICINAL VALUES OF HONEY

Although the general public has not been aware of the value of honey for its medicinal properties, honey has been regarded since the early times as a substance having medicinal values.

Honey Destroys Disease Germs

At the Colorado Agriculture College in Fort Collins, Dr. W. G. Sackett, a bacteriologist, ran some interesting tests.[1] He frankly did not believe that honey would destroy disease bacteria. So in his laboratory he placed various disease germs in a pure honey medium and waited. The results astounded him. Within a few hour's time all of the disease microorganisms died. The typhoid-fever-producing germs died within forty-eight hours. Other similar germs called A and B typhosus, perished after only twenty-four hours. A microorganism found in bowel movements and water which resembled a typhoid bacillus died in five hours. Germs which caused chronic brocho-pneumonia were dead on the fourth day. It was the same with the particular bacteria associated with a number of disease conditions such as peritonitis, pleuritis and suppurative abscesses. Dysentery-producing germs were destroyed in ten hours. Dr. A. P. Sturtevant, bacteriologist in the bureau of Entomology, Washington, D. C. duplicated and substantiated these

---

1. *Bulletin No. 252* published by the experimental station where Dr. Sackett conducted his test, reported the results.

tests. Dr. A. G. Lockhead, working in the Division of Bacteriology in Ottawa Canada, and others varified the work of Dr. Sackett.

Honey-Saturated Bandages for Every Day Use

For day-to-day accidents from cuts, burns or bruises, honey-saturated bandages can be easily applied with amazing results. The honey will aid in healing the wounds. A small bit of honey can be applied on facial blemishes to aid in their healing.

Horehound Coughdrops

Dig out the root of the Horehound plant, wash and snip with heavy shears into small pieces. Place in a saucepan (not aluminum). Cover with water and simmer gently until only about ¼ of the liquid is left. Pour off and repeat this procedure, using the same root pieces. Place the concentrated liquid in a pan. Add same amount of honey. Cook to the hard-crack stage. Drop onto oiled cookie sheet or pull like taffy candy, twist into ropes and cut into pieces with sissors. Keep cool. (from LeArta's *Nature's Medicine Chest*)

Iced Herb Tea

Mix equal parts of sassafras, peppermint or spearmint and alfalfa. (Obtain at most health stores if not available in your own yard or nearby meadows.)

Add 2 tsp of the mixed herbs into saucepan of 1 cup boiling water. Immediately remove from heat, cover tightly and let steep for at least 5-10 minutes. Strain, add honey to taste. Refrigerate. This makes a refreshing and healthful drink. (use warm if desired)

Daily Tonic

One Tbsp of honey, one Tbsp of apple cider vinegar or lemon juice, in a glass of warm water will give energy and it is said will lessen cholesterol.

Honey Gargle

Make a solution of 3 Tbsp of honey, 1 Tbsp warm water. Mix and use as a gargle. It will be soothing and healing.

Pollen

Pollen, like honey, is a gift from the honey bee. Neither can be reproduced by man. Although it is not an essential food for the adult bee it is necessary in the rearing and development of the queen bee.

The magic food properties found in pollen are attracting the attention of athletes throughout the world. The use of bee pollen tablets was pioneered by athletes in Finland as part of a training and feeding program, credited with dramatically increasing the number of Finns among the world's top 100 runners from one in 1967 to 39 in 1972.

The complete composition of pollen is not completely known at present. The future may bring to light new surprises. Its concentration in precious and noble elements makes it a super-food of a biological value. Without exaggeration it can receive the title of a miracle food.

It has been reported that pollen corrects the capillary fragility of delivering women and prevents the menigeal hemorrhages of the baby.

Eating a small amount of dry pollen every day may help immunize an allergic person or those suffering from bronchial asthma.

Pollen enhances metabolism by creating important chain reactions in the digestion system. A teaspoonful for a daily use is sufficient. It becomes a stimulant and may upset you if you take large quantities, especially at the beginning.

CHAPTER 16

THE ADVANTAGES OF BACKYARD BEEKEEPING

Backyard beekeeping is one of the most interesting, beneficial, profitable, and yet undeveloped of man's part-time activities. It is not an expensive activity, and can be a profitable hobby. For example, one hive may produce enough honey in a single season to pay for the initial investment. Where else can you realize such a return with the expenditure of so little time and energy? Financial return aside, the following benefits will more than justify the acquisition of a colony or two.

Nutritional Benefits

Honey is the only concentrated sweet in nature. Outside of of the juices of fruits and vegetables, it was the only sugar known to the ancients. It is mentioned many times in the Bible and its virtues have been proclaimed through the ages. At one time the medical profession considered honey no better than sugar, but there is now irrefutable evidence which recognizes honey as a superior sweet, the only unmanufactured sweet available in commercial quantities. Ordinary sucrose, or what people call white sugar or brown sugar, must be inverted or changed before being absorbed in the human system; honey is already inverted by the bees and goes into the blood stream very quickly. Some late research seems to show that honey is one of the natural foods that is high in vitamins, especially vitamin B1.[1]

Pollination Benefits

It is a well-known fact that bees are absolutely essential for pollination purposes. Farmers often rent bees to help pollinate their crops. People in urban and suburban communities would do well to consider such benefits as they seek to improve the yield of agricultural production on small plots. The owner of bees can sometimes make as much money in a season from rentals as he does from his honey crop.

---

1. *Starting Right With Bees* by A. I. Root, Company, pages 5-6.

Food Storage Benefits

Throughout the ages wise people have stored food, clothing and other essentials in time of plenty as a hedge against lean years. Consider the administration of Joseph in Egypt who brought a nation through a prolonged drought period. Such activity cannot be classified as hoarding. As the bees store surplus honey for use in winter months, so should today's family consider a type of security which is independent of the dollar. The keeping of bees by the family can properly be classified as this type of personal welfare activity. It is possible to purchase a supply of honey for current use and for storage at moderate cost but increasing prices may soon make such a quantity purchase be prohibitive for the average budget. A hive or two of bees, however, properly cared for, would automatically keep the storage shelf in yearly supply. Properly-ripened honey can be stored almost indefinitely with only slight deterioration.

Neighborhood Cooperation Benefits

Backyard beekeeping can either enhance or impair relationships among neighbors. If the suggestions contained in this book are followed, the keeping of bees can contribute to, rather than detract from the welfare and social climate of a neighborhood. The advantages of pollination have already been mentioned. Sharing bees and knowledge can bring neighbors close to each other. Consulting with neighbors before purchasing bees will demonstrate sensitivity for the concerns of those who may be uneasy about a hive of bees in a high density community. The cooperative planting of trees and other crops that provide nectar for the bees could also promote community spirit. It is now commonly known that bees can be safely kept in metropolitan areas on penthouses, rooftops, and in attics.

Educational Benefits

You will be amazed at the interest generated within the family and among neighbors as the workings of the hive are observed and studied. Reading and conversations with others who have bees should precede the acquisition of bees. After the bees are situated on your lot, careful observation and further reading will be a very natural outcome. Your children will be able to give intelligent reports in school, and their circle of friends will increase as they informally discuss their involvement in this family project. Adults, too, will have access to additional conversational topics.

Schools should give serious consideration to a limited bee project in conjunction with a study of the natural sciences. It would not be costly, and would provide motivation for wide reading, the application of mathematics, and other skills and attitudes. A few hives could develop into a year-long project resulting in tremendous learning gains.

Beekeeping by scout troops would be a natural way for boys to earn merit badges and money in addition to the afore-mentioned values. It could become a highlight of scouting experience for many boys in those formative years.

For the hobbyist who has little interest in the financial rewards, backyard beekeeping could provide relaxation and an avenue for creative expression.

CHAPTER 17

QUESTIONS BEFORE INVESTING

If you are considering the possibility of acquiring a colony of bees, you should now be able to answer affirmatively most of the following questions. They will help you to decide whether or not you are ready to invest, either for profit, or for personal satisfaction, or both. A "yes" answer to most of these questions suggests you could make a small investment in neighborhood beekeeping to advantage.

1. *Do I have a place for the bees?* Almost everyone should be able to answer yes to this one. A colony of bees may be kept on a city lot, on the flat roof of a garage, in a work shop, or in your attic. Screening may be necessary to change the bee flight pattern in order to avoid their contact with people.

2. *Is it legal for me to keep a colony of bees?* In some cities there are restrictive ordinances which should be investigated. If keeping a hive of bees on your property is against the ordinances, you may be able to convince the city officials that you are entitled to a variance as to provide benefits without harming or harrassing your neighbors.

3. *Do others in my area keep bees?* You may be surprised to discover that your neighbor just beyond the high board fence has

a hive of bees. You should contact him to find out whether the project is successful. You may also be able to pick some valuable pointers on how to operate your bees. Most beekeepers are very friendly and willing to exchange experiences and information and to give you a hand in getting started.

4. *Will the bees produce enough honey to make it worthwhile?* In most areas the sources of nectar will justify the keeping of a limited number of bees. The bees will fly a radius of several miles if necessary to obtain nectar. The present high price of honey would make this an extremely tempting project.

5. *Will my neighbors approve?* If they can be assured it can be done without danger to them or their children, there should be no objection. In general, the more you can bring your neighbors into your project by sharing some honey and your experiences, the more likely they are to cooperate with you.

6. *Are some of my neighbors interested in helping with the project?* You may find that getting others involved will create a much closer spirit of unity in your neighborhood, especially if others are financially involved with you. This is a skill—almost a lost one—which might be most valuable in the event of stressful situations such as food shortages, transportation tie-ups, etc.

7. *Can I afford it?* Even with rapidly increasing prices, it is very possible that you may be able to recover your investment the first year. After the first season, you can generally divide your hives, making two from one. Start small while gaining your experience.

Some of these questions could be answered by contacting the State or County Bee Inspector or a friendly nearby apiarist, or you might find a neighbor who has taken up beekeeping as a hobby who would gladly give needed information. He might sell you a hive of bees and even volunteer some assistance.

CHAPTER 18

TERMS YOU NEED TO KNOW

Before beginning the study of any specialized area it is important to have at least a limited knowledge of the basic and most commonly used terms. For this reason, a brief listing of beekeeping terms is provided so the remaining chapters will be more meaningful.

APIARIST
A commercial beekeeper.

APIARY
A number of colonies of bees in one location.

BEE VEIL
A protective net covering worn with a hat.

BEE WIRE
A fine wire used in the frames to support a newly-made comb.

BEEHIVE
A box of uniform size and measurements. A standard 10-frame hive today measures 16" in width, 20" in length and 9 5/8" in depth.

BOTTOM BOX
The beginning of a hive structure with frames; a bottom board and lid.

BROOD
One or more capped-over cells, each containing the pupa; darker in color than capped honey.

BROOD NEST
One or more frames containing brood.

CELL
A hexagonal structure of wax made by the bees.

COLONY
A hive box, a number of worker bees, a queen, and one or more drones. The colony of bees may be one or more stories high.

CHUNK HONEY	A delicacy. The honeycomb taken by the capping knife or spoon method could be put in bottles or containers and sold as chunk honey. Large chunks of sealed comb with some liquid honey poured in makes an attractive package and usually sells for double the price of extracted honey.
COMB	A multiplied number of hexagonal cells.
CRYSTALLIZATION OR GRANULATION	Hard or candied honey. A sign of pure undiluted honey.
DRONE	A male bee with the primary responsibility of mating with the queen. It has no stinger.
EMBEDDER	A tool used to embed wire into the foundation (manual or electric).
FEEDER	A long narrow tray or can for supplemental feeding of honey or sugar syrup.
FILTERING	Process by which honey has been filtered or finely strained.
FOUL BROOD	A disease of the larvae.
FOUNDATION	A thin sheet of wax or plastic stamped the exact size and shape of a hexagonal cell centered in the bee frame and fastened to the top bar reinforced by embeded parallel wires. (Pre-wired foundations are also available)
HONEY	A pure natural sweet processed by the honeybee, gathered by bees from the nectar of flowers. It is a non-carrier of germs; therefore, it will never contribute to the development of crippling diseases. Used on burns, wounds, ulcers, dressing following operations as an antiseptic and antibiotic.

HONEY-COMB — Structure of wax made by bees consisting of a mass of hexagon cells to hold honey or larvae.

HONEY-DEW — A sweet secretion found on leaves of certain plants, not always desirable for human food or for storage by the bees.

HONEY-FLOW — An abundance of nectar.

HONEY SACK — The honey-bag organ in the bee in which honey is prepared.

LARVA — The stage of life about three days after the egg is deposited.

NECTAR — A sweet liquid secreted by plants, the chief source of honey for bees.

NUCLEUS ("NUKE") — A method of making increase by placing within a new hive two or more frames of hatching brood with a fully developed queen cell.

NURSE BEE — A newly-hatched bee assigned to hive work.

POLLEN — Colored, dust-like material found on anthers of flowers and gathered by bees, a needed protein for the development of larva.

PROPOLIS — A sticky substance gathered from buds of trees and used by bees as a cement.

PUPA — The second stage of life after the larva, showing form.

QUEEN BEE — The mother and ruler of the hive. The quality of the queen largely determines the strength of the hive. Some times a queen becomes a drone-layer because of inability to produce fertile eggs, because of injury or other causes.

QUEEN EXCLUDER — A wire mesh with openings large enough for the passage of worker bees, but not the queen.

RAW HONEY — Extracted honey which has not been highly strained or heated therefore pollen grains and wax particles are retained.

RIPE HONEY — Filled cells in the honey comb that have been fanned with the bee's wings and capped. Ripe honey does not ferment or spoil. It may granulate but this is only evidence of good ripened honey. Long periods of storage may turn the honey dark and it may lose enzymes, diastase other digestive properties, but does not spoil.

ROYAL JELLY — A special preparation by the worker bee for the development of a queen bee.

SMOKER (bellows) — A small bellows and can with snout used to sedate bees with smoke.

A SMOKER

SUPER   Additional hive body above the brood nest.

SWARM   A number of bees settled together, as in a hive. A body of honeybees which emigrates from a hive and flies off together, under the direction of a queen, to start a new colony.

TERRAMYCIN   A drug used in the control of foul brood.

WAX MOTH   A small gray moth which deposits its eggs on the combs and frames of uninhabited or unprotected hives. When the pupa hatches, it feeds on the wax and wood reducing the hive to complete ruin Prevention: place mothballs in the hives or use sulphur fumes.

WORKER BEE   An undeveloped queen with the burden of gathering nectar and honey, producing wax, building the comb, protecting the hive, controlling temperature, etc. (sometimes called a field bee)

CHAPTER 19

COMMON QUESTIONS AND ANSWERS
ABOUT BEES AND HONEY

Included on the next few pages are some of the questions most commonly asked by beginning beekeepers.

Q. *How are bees obtained?*

A. From a neighbor, from stray swarms, or by purchasing of bees

Q. *What is a good location for bees?*

A. Bees can be kept with little danger of bee stings in most backyards, on garage roofs, in attics, or on city roofs within range of nectar-bearing plants.

Q. *How many bees are in a hive?*

A. 50,000 is considered an normal strength for a colony of bees, but they may increase to over 150,000. During winter months the count may drop much lower.

Q. *What is the length of time required to hatch a bee?*

A. Fourteen days for a queen; 21 days for workers and for drones.

Q. How long does a bee live?

A. During the honey season, worker bees live about six weeks and in the winter about six months. With all their summer flying their wings wear out and the bees are lost in flight. An average queen will live about three years. Drones are eliminated at the end of the season by not being allowed to eat the honey reserves.

Q. Why are there dead bees in front of the hive?

A. It is normal for bees to remove dead bees from the hive that have died during the winter. However, a large number of dead bees in front of the hive in the summer may indicate poisoning from insecticides and is cause for alarm.

Q. Why do bees sometimes cluster on the outside of the hive?

A. It is evidenced they are crowded for room and if not given more space swarming may occur. If bees are over smoked they sometimes cluster outside for fresh air.

Q. Why do bees swarm?

A. It is mainly to perpetuate the race. When the hive becomes crowded with bees and honey, swarming is induced, which is a time of bustle and excitement. Not all the bees leave the hive when swarming; the old bees are left to carry on the hive strength.

Q. Why do bees sting?

A. Bees seldom sting except to protect their home against invasion by man, beast, or other bees. To sting costs them their lives. Exceptions to the rule: (1) if stepped on by bare feet; (2) if accidentally pinched when handling the combs; (3) if swatted at in fear.

Q. Does the queen sting?

A. The queen bee has a smooth stinger but never uses it except in combat with a rival queen.

COMMON QUESTIONS, ANSWERS ABOUT BEES AND HONEY

Q.   What should be done for a bee sting?

A.   See chapter 22, page 105.

Q.   Is a bee sting dangerous?

A.   It is not serious except to a few individuals who are allergic to this type of poison. A bee sting on the eyelid will cause no damage to the eye. The swelling thus caused will recede after about the third day.

Q.   Why do bees appear at your screen door?

A.   Only to obtain food when there is no nectar in the fields. They will enter a home through an open door when fruit is being canned or when someone is cooking with sugar or honey syrup. They do not come to sting.

Q.   Where do bees get honey?

A.   From the nectar of blossoms, such as flowers, fruit trees, alfalfa. clover.

Q.   How far will bees fly for honey?

A.   Normally bees fly one or two miles, but they will go as far as eight miles in search of nectar.

Q.   What is the normal output of honey per colony?

A.   Normally one could hope to extract about 40-50 pounds a year. Different localities have different yields.

Q.   What can honey be used for?

A.   Honey is a basic food that can be used as any other sweetener. It also has medicinal value such as healing burns and cuts.

Q.   Is honey better than sugar?

A.   Yes. Honey contains vitamins, minerals and digestive properties not to be found in other sweeteners.

Q.   Can you cook or bake with honey?

A.   Yes, but you should cook first and then add the honey; otherwise you destroy some of the honey's essential food elements. Use lesser amounts of honey than sugar.

Q.   How do you liquify granulated honey?

A.   Place the honey container in a larger vessel filled with water, then heat to a simmering temperature. Provide space between the two containers at the bottom to avoid overheating. Honey melted slowly will retain its fragrance and natural food values.

CHAPTER 20

SEASONAL COLONY ACTIVITY AND INDIVIDUAL BEE DEVELOPMENT[1]

Colony Activity

Basically a colony of honey bees comprises a cluster of several thousand workers (sexually immature females), a queen (a sexually developed female) and depending on the colony population, an undetermined number of drones (sexually developed males). A colony normally has only one queen, whose sole function is egg laying and not colony government. The bees cluster loosely over several wax combs, the cells of which are used for the storage of honey (carbohydrate food) and pollen (protein food) and for the rearing of young bees to replace old adults.

The activities of a colony vary with the seasons. The period from September to December might be considered the beginning of a new year for a colony of honey bees. The condition and activity of the colony at this time of year to a great degree affect its prosperity for the next year. In the fall the night temperatures decrease and the days shorten. Plant growth, nectar secretion and flower characteristics respond in various ways that affect the activity of the bee colony.

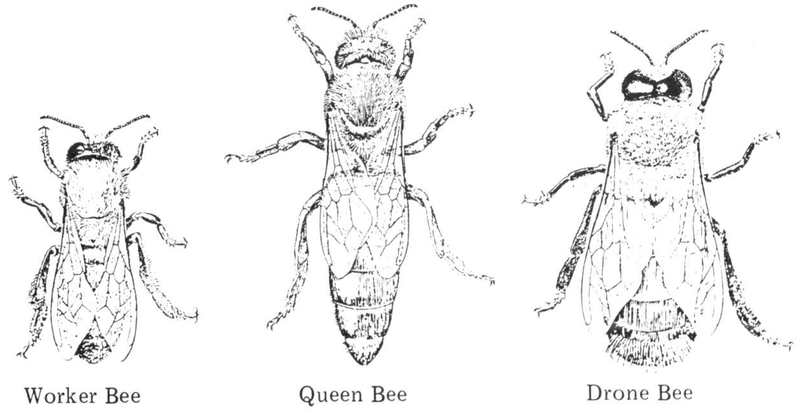

| Worker Bee | Queen Bee | Drone Bee |

---

1. A portion of this chapter is adapted from *Starting Right With Bees*, The A. I. Root Company, p. 5.

Early Stages in Bee Development

The queen lays small, elongated, white eggs, one attached to the base of each cell. The egg remains in an upright position for 3 days, then a larva hatches.

The larva, a small white grub, is mass fed royal jelly by nurse bees for the first 2 days to such an extent that it literally floats in this food. For the next 4 days the worker larva is fed a less nutritious food at regular intervals (progressive feeding) by a different group of nurse bees. During these feeding periods it grows rapidly to several times its size at hatching and soon occupies the greater part of its cell. Its cell is capped on the ninth day after the egg is laid. No feeding occurs after the cell is capped.

The next 12 days are spent in the capped cell in prepupal and pupal stage. Following the fifth or last larval molt, which occurs 2 or 3 days after the cell is capped, the distinct adult body parts appear such as legs, antennae, wings, mouth parts, head, thorax, abdomen and eyes. Initially the pupa is white and extremely soft textured, but as it grows to emerge from the cell as an adult, it changes from white to a darker gray and finally to its adult color. Hair develops on the various body areas and on the day before emergence it completes its sixth and final molt. On the 21st day the young worker bee cuts its way through the capping and crawls out of the cell as a physically adult honey bee.

Activities of Young Bees

The activity of a honey bee within the colony varies according to its age and the development of its internal glands. During the first 3 days following emergence, it aids in cleaning the next area. From the fourth to seventh day after emergence it feeds older larvae with mixtures of honey and pollen. About the seventh day after emergence its pharyngeal glands become developed and it produces royal jelly. As such a nurse bee it mass feeds the young larvae or queen until it is 10 to 13 days old.

About the 12th day of its life the wax glands, located between the fourth and seventh segments on the undersurface of the bee's abdomen, develop. Through some still unknown body process honey is eaten and converted into beeswax, which is secreted as small scales between these mandibular actions into cell walls in the construction of combs. When the bee is about a week old, it takes short orientation flights during the afternoon to memorize its home location. Various activities such as the conversion of nectar into honey, packing pollen pellets in the cells, handling water brought in for air-conditioning, fanning for ventilation and guarding the entrance are performed until about the 21st day, after which it begins foraging for food.

The Adult Bee

Adult honey bee workers live about 6 weeks during the peak activity periods of late spring, summer and early fall. Most of them die in the field. Those that die in the nest are carried out some distance from the entrance and dropped to the ground.

While it is true that the queen does not govern the hive, if by injury or other reasons she is removed from her home, a state of mourning ensues. The bees have a different hum; wax building, except for building queen cells, ceases; field trips for honey are lessened; pollen gathering is temporarily discontinued. Movement within the hive denotes confusion and lack of purpose. The worker bees proceed to build a number of queen cells by enlarging a cell containing young larvae. The colony then waits out a 14 to 16 day period when a new queen emerges and full hive activity is resumed.

The adult bee lives strictly on a mono diet—honey—and without honey, it is said, they cannot live more than four hours. Obtaining this precious food does not come easily; a bee would have to fly a distance of more than three times around the world to gather one pound of honey.

The secret, then, of a bumper honey crop is to have a large number of bees—up to 150,000 strong in a single hive.

With a consistent pattern of flying in search of nectar the honey bee substantially increases the yield of fruits, vegetables and other crops.

A drone does not die of old age but is banished from the hive when the season is ended. His flight speed is up to 90 miles per hour.

CHAPTER 21

GETTING STARTED IN BACKYARD BEEKEEPING

Equipment Needed

It is suggested that in getting started with this hobby or vocation—especially if one is a novice—it is better to acquire only one colony of bees. This means one hive, consisting of a box or super with its frames, bottom and top. Later in the first season you will need another super complete with frames; and if your bees are in a heavy nectar-bearing area, you may need even a third super or more.

You may buy your first equipment new from bee supply houses and then buy a package of bees. Perhaps a better way would be to buy an existing colony of bees from a neighboring beekeeper or a commercial beekeeper. He may not have a complete hive for sale but might be prevailed upon to sell you a "nuke," which is 2 or more frames of hatching brood and bees with a developed queen cell. Those frames could be transferred to your own equipment. This "nuke" should not be removed to your location until the queen is hatched and is mated perhaps a week or more. When mated, she will begin to lay eggs which are visible in the cells.

You might advertise for a hive in your market bulletin or local paper. By calling your county bee inspector, you may be able to get a list of the beekeepers near you.

A good source for bees alone, without having to pay for them, is to contact your local police department. They often receive calls from a frightened housewife about a swarm of bees temporarily alighting in their shrubbery, trees or attic. The swarming season is generally in the spring and early summer but may occur even later. The author and five of his immediate family have each obtained all the bees they have wanted as hobbyists, by using the swarm capturing methods described later in this chapter.

Packaged Bees and Queens

A 2, 3, or 4 pound package of bees may be ordered with a mated queen. The Italian strain of bees is most common and is of gentle nature, less apt to sting. They are also good honey producers. The other well-known strains are Carneolian and Caucasian. The

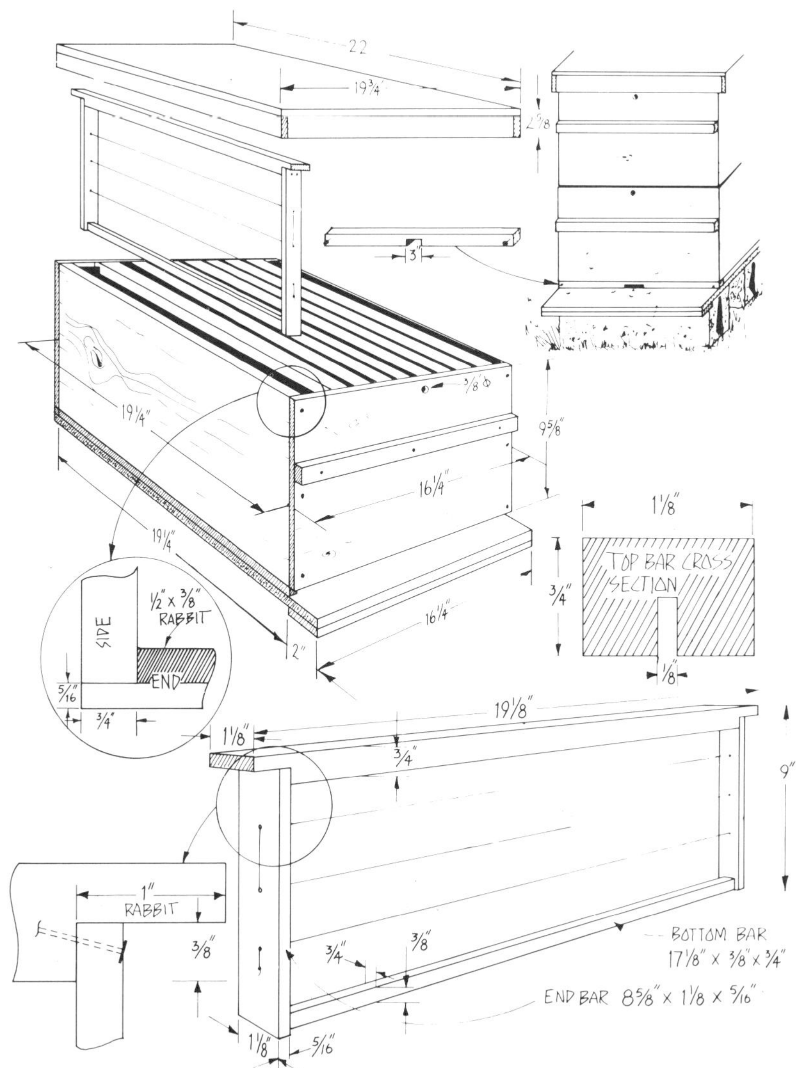

DIAGRAM FOR CONSTRUCTING A BEEHIVE

Caucasians are even more gentle than the Italian and are ideal for backyard beekeeping.

The small black bee from Germany is of a nervous nature and more inclined to sting. A good cross for gentleness and for increased honey-gathering capacity is a Carneolian Queen crossed with a Caucasian or Italian drone. Since the lifetime of the bee is from six weeks to six months with the new improved queen, you can completely change your strain of bees by simply changing to a better queen.

A package of bees comes to you by air in a screened box with a can of feed syrup installed. They may be ordered with or without a mated queen. If the package includes a queen, transfer the queen cage immediately into the new hive. You may want to order just bees to add to an existing colony that has been weakened by a hard winter in order to get ready more quickly for the honey flow season.

Helmet & Bee Veil

Bee Suit

PROTECTIVE CLOTHING FOR THE BEEKEEPER

When the package arrives, you should be ready with an empty hive and foundation frame or combs placed on your selected location. Transfer the bees immediately on arrival. *Before transferring, however, be sure to read the instructions that will be sent with the package.*

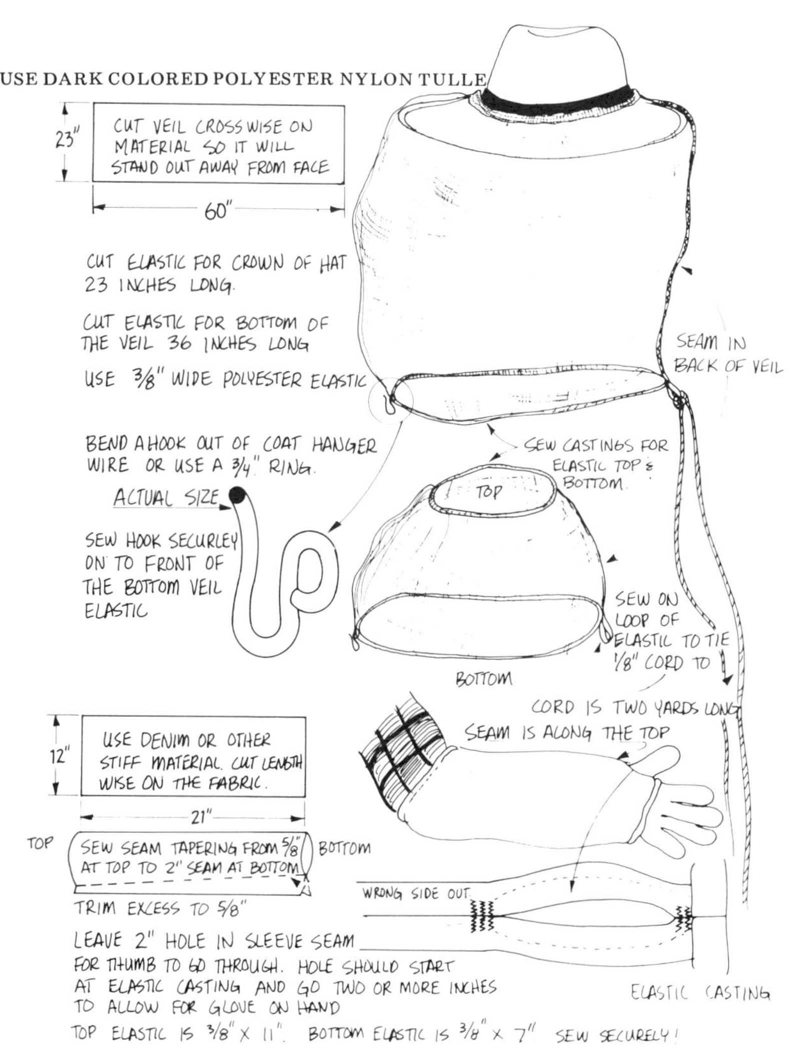

INSTRUCTIONS FOR MAKING PROTECTIVE CLOTHING

How To Transfer Bees From Package To Hive

Wear your veil and have your bee smoker ready, even though it is not generally needed at this phase of the operation. Package bees are seldom unruly and rarely sting. Gently remove the feeder can then spray the bees thoroughly with a sugar syrup before inverting the cage into the hive, having removed sufficient frames to accommodate it. If the bees are slow in exiting, raise the shipping cage a few inches while it is still in the hive and gently shake the bees into the hive.

Caged bees in transit seem confused when dumped into a new home and are sometimes slow to orient to their new location. There may be an unusual amount of flight and milling around at first, but the bees are not angry, just confused. This is no problem unless a number of packages are introduced into adjacent hives at the same time. In this case there may be some drifting of the bees from one hive to another. This can be minimized by having some object, such as a bush or a large bucket, board, limb, etc., between the hives to aid the bees in establishing themselves in their new location. As the bees make their first home orientation flight, it is not unlike a person finding his new home in a brand new tract of identical homes! It is well to be prepared with a standard feeder in the event cloudy, cold or rainy days prevent the bees from gathering nectar, since they will have no reserve; under such conditions starvation can occur.

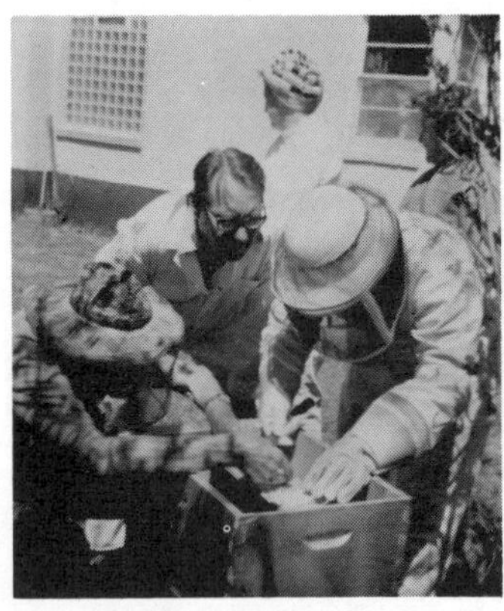

If a queenless package is used to strengthen a weak colony, be sure to smoke the existing one quite heavily and then gently shake the bees from the package onto the top of the opened existing colony immediately. The heavy smoking will minimize the fighting which generally occurs when trying to put two groups of bees together.

How To Capture A Swarm Of Bees

Swarming is a two-stage process. The bees first cluster in a tree, bush, shrub, or on a fence post within close proximity of their original hive—sometimes only a few yards away, sometimes farther. This is a temporary location for from a few hours to a day or longer. The scout bees leave the temporary home to search out a permanent home, which could possibly be many miles away in relatively inaccessible places, such as a hollow tree, a crevice in the rocks, etc. Therefore, you had better capture them while they are in their temporary location. This means you should be prepared with your equipment in advance. However, in emergencies you can capture a swarm in a cardboard box and later transfer it to a permanent hive.

CAPTURING THE FIRST SWARM

SWARM IN A TREE

The Hive Has Been Lowered To The Ground
The Remaining Bees Are Being Smoked From The Limb

An empty hive may be placed in one of your own trees, baited with oil of anise and powdered cumin. This mixture smeared on a chip of wood and dropped between the frames will attract the scout bee who may bring a swarm to you without charge. If you already have a colony or two of bees and want to capture your own swarm, the baited box should not be put adjacent to the existing hives, as the swarm will pass it by. They always seem to settle at least several feet and preferably several yards off the ground. For this reason your swarm box should be placed in a tree, on a roof, etc., to be noticed by the scout bees.

Any pattern or method of capturing a swarm is subject to failure since the bees are unpredictable in their swarming habits. For example, one beginner having placed a swarm box in a nearby bush was surprised to have the swarm bypass the box and settle on an over-turned fruit basket in the carport.

If a regular bottom box is not at hand, a cardboard carton (preferably of standard dimensions to receive two or three frames) may prove even more advantageous because of its light weight—especially if the bees are clustered high in a tree. It will be easier to carry a lightweight box when climbing a tall ladder or to use a net fastened to a long pole. Another alternative is a specially made swarm box with a large funnel provided by supply houses.

With a sudden jar of the limb, the main cluster will fall into the box or net and then should be quickly brought down and poured into the regular hive box. Since the queen is in the center of the cluster, it is most likely you will have the queen in the box. Some of the bees will re-cluster on the limb where they can be shaken into the box or net and brought down as in the first operation. It may be necessary to repeat that two or three times. It sometimes helps to use the smoker, driving this second cluster from the limb. If the queen is in the box, the bees will eventually give up the limb and settle down; then they can be carried away to another location.

In swarming time, bees are most unpredictable. On one occasion a large swarm flying over a construction area settled on a pile of earth freshly excavated from a trench. The workmen nearby threw down shovels and tools and ran for shelter. This caused great excitement. The foreman of a carpenter gang detached one of his men (the author) to capture the bees. A cardboard box was improvised, an entrance cut so that the flap could be closed after they were in the box. The box was placed near the main cluster and with my bee smoker which happened to be in the car, the bees were gently driven into the box through the entrance. In a short time the box was closed and carried away and the frightened men returned to work.

TRANSFERRING A HUGE SWARM FROM AN OCCUPIED FRAME HOUSE

Bees that have clustered on posts or some other inaccessible place can be gently scooped into the box with the hands or if available with a regular dustpan or similar scoop. If the queen is captured, some of the bees still re-clustering on the post can be dissuaded by the use of a smoker or brushed carefully from the post.

It is always well to go prepared with a bee veil, smoker, or spray can (canned smoke) in the event the bees have settled in a difficult location.

A makeshift smoker could be made by wrapping a ball of burlap or other material around the end of a short stick. When lighted it will serve as a smudge to be fanned over the bees. Ordinarily a smoker is not necessary in catching a swarm of bees, because usually they will permit you to help them in a hive without resistance.

Transferring the Bees to a Location

The swarm, if carried in a cardboard box, can then be carried to a location. The regular box should later be placed on the exact spot of the cardboard box. When bees orient to a certain spot, they cannot be moved even short distances during the summer flying season. Even three or four feet will cause the loss of field bees.

In case it becomes necessary to move the bees to some other spot in the backyard, the bees could be taken to a temporary location, at least a half mile away and left there for a few days; then returned to any desired location.

Where to Locate the Bees

The proper location for the bees is of vital importance from the point of convenience, protection from vandalism, safety of family members and neighbors, and also for the welfare of the hive if located in colder climes.

Use wire fencing of your choice to enclose a 6' x 8' space in a suitable corner where the bees will be assured the winter sun. A flowered climbing plant on the fence would add to efficiency and make the appearance more attractive.

A BACKYARD LOCATION FOR A BEE COLONY

After deciding on some corner of the lack lot, make an enclosure approximately 6 x 8 ft. square and not less than 6 ft. high with a gate; the material may be grape stakes, a tight board fence or a light framework covered with chicken wire on which vine could be trained. The enclosure may be constructed of natual landscaping, such as a hedge of shrubs or trees.

The provision of an enclosure will apply to any location, the orchard, on school campus, the farm, or the front patio. This also provides safety for small children, who in the absence of adults may approach the hive with sticks or rocks, provoking the bees to anger. It might also deter attempts from older boys to invade the hive, resulting in bee stings, doctor bills, lawsuits and the loss of friendship. Such an experience actually happened in Provo, Utah.

These arrangements provide an improved flight path with little conflict to garden traffic. The most important factor in cold climates is that the hive should be warmed from the winter sun reflections. While bees are semi-dormant in winter, they must have cleansing flights during the winter months. The reason for this is that honey bees never excrete in the hive (a happy thought for those who love honey). If a shaded yard prevents this flight the bees may contact dysentery and die. On a warm sunny day in January, the bees will leave the cluster for a brief flight evidenced by brown spots on the white snow. In northern sections subject to winter and spring winds, a sheltered spot affording winter sunshine is of vital importance.

If fencing material is not available, it may be that trees and natural shrubs will provide a suitable location. In any event, the proper place for a bee yard is of utmost importance. If these provisions are neglected, your project could be a failure.

For those who have acreage away from home, or can arrange for a bee location, the problem of providing an enclosure will be lessened; however, many of the rules will still apply regarding the safety, protection and welfare of the bees. This more remote location could accommodate expansion from one or more colonies in the backyard to a small apiary and may become more than a hobby by providing a source of income and the security of a basic food for current use and storage.

Moving a Hive of Bees

The hive should be moved only when the majority of bees are in the hive. The occurs near night fall in warm weather. In times of wet or cold weather, the bees can be moved in daylight. If you move them when the worker bees are out, you not only lose the worker bees, but they may become a nuisance when they return and find their home has disappeared.

Since bees will come out of their hive in an angry mood when disturbed in the moving process, be sure that you can close up all fly holes and fasten bottom boards and supers together so they do not shift apart in the move—especially if you are carrying them in a station wagon! Bees do not fly much at night but they certainly crawl—in embarrasssing places.

When you place the newly acquired hive on its permanent location, be sure to unplug the entrance or the bees may smother in a short time, especially in hot weather.

CHAPTER 22

RULES FOR WORKING WITH BEES

Rules of the bees are strict and must be observed. Remember, bees cannot be domesticated and will never become your friends in the same sense that you can win affection from your prize horse or pet dog. They will, however, permit a good working relationship with little danger of a sting. If you follow these simple rules you can work safely and happily with these tiny tireless employees who are unaffected by labor unions or time clocks.

A hive inhabited by 50,000 to 100,000 bees will work for you from daylight until dark through the warm summer days, bringing their sweetness from the flowers, turning it into a food that will need no additives, no drying process or refrigeration, but will keep indefinitely as one of your best basic foods. Here then are the rules:

Rule 1: *Smoke The Bees*

This rule applies on all occasions, especially for beginners. Smoke acts as a tranquilizer as well as a means of driving the bees from the area to be worked. First, apply a few moderate puffs at the entrance to drive back the guards. Then, as the inner cover or burlap is carefully lifted from one corner of the hive, apply smoke in the same manner.

Before removing the first comb, provide free space on both sides by sliding adjacent combs away and moving frames closer together. Avoid scuffing or angering the bees as you withdraw the first frame, since there is a possible risk of injuring the queen. Apply smoke only if necessary to keep the bees quiet. Do not jar or drop the super when adding it to the hive.

Rule 2: *Never Open The Hive Late In The Evening
Or Early In The Morning*

Most of the worker bees will have returned from the fields, making inspection difficult. The bees will be touchy and offer resistance. The hours of 10:00 a.m. to 5:00 p.m. are usually safest. The bees will let you know if you make a mistake!

Rule 3: *Do Not Work The Hive On Dark, Cloudy, Rainy Days*

The restless full field force and all the guards will make you very unwelcome!!

Rule 4: *Approach The Hive With Caution*

This is especially important when supers are filled with capped-over honey.

Rule 5: *Working The Hive In Cool Summer Weather After The
Honey Flow Is Over May Require Increased Smoking With
The Application Of All Safety Rules Of Precaution*

Under such conditions the bee colony's guard is doubled to prevent the stealing their summer's work and winter's supply.

Rule 6: *Study Carefully The Temper And Behavior Of Your
Bees Under All Conditions*

Treat them as your friends. Learn their sign language. You may have a pleasant relationship or you may get stung. It is up to you.

Rule 7: *Do Not Be Frightened If A Bee Happens To Gain
Entrance Inside Your Veil*

In most cases it will be so busy trying to find a way out that it will have no intention of stinging. Retreat quietly where you can be safe from other bees. Then remove or lift your veil and release the bee. Approaching the bees with fear may attract their keen sense of smell and invite them to combat. Relationship between the master and the bee can become so congenial that the bee inside the veil, under the sleeve, or up the pants leg, will seldom sting.

Compare the veteran, who works in safety without veil or gloves, to the beginner: veiled, with gloves and pant legs tied, a frightened look on his face as a bee circles, not in anger but ordinary flight, swatting as he backs away with unfriendly expressions such as "You little begger, get lost!" The following incident is an example of what can happen.

A number of years ago in Spring City, Utah, an attorney decided to try backyard beekeeping. A few hives had been secured. As the summer advanced, the attorney decided to taste some comb honey. With no knowledge of bee behavior and poorly veiled, he opened the hive without a smoker and proceeded to remove a comb. The bees charged and stung him. The attorney retreated, inflicted with many stings and filled with anger. He returned, dressed to ward off any bee attack, his bee veil well fastened, pant legs tied at the bottom and gloves and sleeves reinforced by stuffing rags into any possible opening. He approached the hive, giving it a kick. He then exclaimed, "You little sons of bees, if fight is what you want, fight is what you'll get!!!" At that he kicked over the hive. The bees attacked in great numbers where he least expected, crawling through his shirt, between buttons and under his well-fastened bee veil. They even bored under the belt of his pants and found entrance. The attorney once again was forced to run for cover.

This extreme example need not be frightening. It will never happen to you if proper procedures are used and if you conduct yourself with a more friendly attitude toward the honey bee. Remember Abraham Lincoln's motto, "If you want to gather honey, don't kick over the hive."

When a bee is very angry he flies straight for the target and is as swift as a rocket. The bee that buzzes around the head will seldom sting unless you fight back. Don't evidence fright—just stand perfectly still.

Occasionally one or two older bees of the hive will continue to seek revenge. In such cases, it is better to eliminate them. This can be done with a small paddle or hand-full of dry twigs. As the bee buzzes around the head, a swift motion with a paddle will invite the bee to combat in which it can be smacked down. Another secret: when the bee is buzzing around for an attack, remove your hat and invite the bee to land in your hair. The bee will crawl forward through the hair, thus not able to sting until it gets turned around. This will give you ample time to either smack or pinch the bee to death.

It is true that some strains of bees are ill-tempered and will go searching for trouble. Much, however has been done and will continue to be done to overcome this by introducing queens from gentler strains.

On one occasion three of my grandsons were preparing a spot on a side hill for additional colonies which were to be added to the original apiary. Before leaving I approached the nearest hive for the purpose of gently lifting it to determine the hive weight. The bees rushed out in great numbers, attacking all of us in fury, stinging without reservation, until we were compelled to retreat. (A gentler strain of bees would never object to this careful operation.) The following day I returned properly dressed and with smoker. After a few lively puffs at the entrance and then smoking by degrees as the lid was gradually removed, the bees became manageable and offered no resistance. A comb of honey was removed and given to the boys as a peace offering.

Your bees can either be kept friendly or, by wrong procedures, can be stirred to anger until they attack man or beast, in which case you will have to destroy or move them.

Rule 8: *"Be Prepared And Fear Not"*

Under all conditions, and for all occasions, apply this over-all rule. This will be the key to your success or failure.

The execution of rules and general procedures will vary somewhat in areas with different climatic conditions and sources of nectar. But even after obeying all these rules, there may be an occasional bee sting.

If A Bee Stings

When a bee inflicts it's barbed stinger, it cannot be recovered, Part of the bee's body is left attached to the stinger, including a small bag of poison which continues with a pumping action to inject a full measure of its contents. This costs the bee its life.

If you are stung by a bee, your first act should be to immediately scrape away the poisonous bag with a fingernail, hive tool, or whatever is available. Don't pull out the stinger: the squeezing motion forces more poison into the blood stream. Suck out the poison before it is carried into the blood.

The broad Plantain leaf may help reduce the pain and future swelling. Crush the leaves and rub the juice onto the injured spot. A tea made from Plantain may be used if the poison is in the body, or one can chew the leaves while drinking water. Lobelia and comfrey are also effective in neutralizing the poison. The application of household Ammonia will prevent swelling. Some have used honey.

The poison of a bee sting is alkaline in nature, so if the victim's blood is acid, a sting may be serious. In this case, consult a doctor immediately. If the person is not allergic, the swelling and soreness, which lasts two or three days, will not be serious. Repeated stings in the early season sets up an immunity and while pain from the sting does not lessen, swelling will not occur. If you wish to speed up immunity, consult your doctor for shots.

If you want your experience in beekeeping to be pleasant, use the proper methods. It is now your inning, *and if you play by the rules of the bees:* applying smoke moderately, and keeping the bees under control at all times, the experience will be pleasant and filled with interest.

CHAPTER 23

THE FIRST SEASON

Preliminary Inspection

If you have acquired package bees, a stray swarm, or a nucleus, you will need to inspect them to determine the level of general hive activity. If all the foundation frames or combs have been drawn out by the bees and are well filled with honey and brood, and the hive weighs approximately 40 pounds, it is time to add the first super. If the hive is too far below this condition, give them more time before adding a super. Oversupering is not a good practice because it seems to discourage the bees.

Adding The First Super

A common practice is to intersperse well-filled combs with foundation frames. This is done by removing two or more frames from the bottom box to be interspersed with the foundation frames in the first super above, replacing of course, the combs taken from the bottom box with foundation frames. This method can be followed as you add the second, third or fourth super. Where a queen excluder is not used, brood rearing can largely be restricted to the first super by placing well-filled combs of honey in the center of the added super because the queen prefers not to deposit eggs in outside combs. It is also the common practice of bees to store honey above the brood nest. Another method is to remove the filled bottom box from the bottom board, replacing it with a super foundation of frames, and then replacing the original bottom box of filled combs on top as a super. It seems natural for bees to build downward. They are quite adverse to

ADDING THE SUPER

going from a box or brood to an upper story with nothing but frames of foundation. This method can be repeated as additional supers are added. In all cases, frames should be evenly spaced about 3/8" apart to avoid burr combs from being built between them. When you have purchased package bees or nuclei, swarming may not occur until the second year.

Natural Swarming

Why do bees swarm? This question was presented to me by my daughter Lyona as we were writing this book and it struck me with great force and impact. An attempt to answer this question involved bee behavior so filled with interest and magic organization that it can never be fully understood.

Swarming is nature's way of reproducing, its guarantee against extinction. It is the prime purpose of hive activity. Whence comes the high order of instinct that motivates the bee in the many details involved in this exciting exodus? When should they swarm? Why does the queen go forth with a determined number of young field bees? Why do not all the bees swarm, leaving the hive empty? What impels the queen to set the day and hour or causes her to postpone the flight because of rain or bad weather? May I repeat again, only God could create the magic of the hive.

Natural swarming may occur when the queen has produced

eight or more full frames of hatching brood. Other factors contributing to swarming fever include need for more room, a sufficient honey flow to sustain the swarm while building a new home, and the capability of producing a sufficient supply of wax to build new combs.

In preparation for swarming, it is the nature of the old queen to supply the home with from one to fifteen unsealed queen cells. When the first queen cell is sealed, swarming will occur. The queen sounds her bugle. The younger bees respond and rush to the exit as if their home were on fire. What determines the age and number of bees leaving with the old queen? The departing swarm displays new vigor and will power in building new combs and searching the fields for honey. The old bees remain to witness the royal battle of the young emerging queens, out of which only the strongest one survives. The first queen to hatch immediately proceeds to open all unhatched queen cells and pierces the helpless victims to their death.

For the beginner it is best to permit natural swarming the first year. The uncertainty of the time when the swarm will emerge can be lessened by weekly inspections of the hive for queen cells in progress, etc. When it is expected, take time to observe this phenomenon.

It will be an exciting moment when you witness this rushing from the hive. You may see the forming of a small cluster in a nearby tree or shrub. This will be the queen surrounded by her scouts. Watch how fast the cluster grows.

Why does the swarm choose to cluster on a nearby bush or tree as if waiting for the owner (master) to provide a new home? In reality this is a temporary location while scouts search for permanent quarters. The first bees to cluster with vise-gripped claws sustain not only their own weight but the weight of many others.

The thrill increases when the cluster is gently shaken in front of the new hive box or directly into the open box as the situation requires. In a short time they settle down and continue their lives in normal hive activity.

CHAPTER 24

EXTRACTING THE HONEY

You may choose to extract honey when your first super is filled, or you may defer until the honey flow is over. In this case, extra supers may be required.

How To Take Honeycombs For Extracting

When the time comes to take honey for extracting, the colony will be at its peak in numbers, perhaps totaling as many as 150,-000 bees. They will be reluctant to part with their stores of honey, especially if the honey flow is over. It may be necessary to increase the amount of smoking in order to drive the bees from the sealed honey combs.

TAKING HONEY
FOR EXTRACTING

UNCAPPING

A 4-inch wide paint brush or a wallpaper paste brush should be used with gentle movements. Dipping it in a pail of water will keep the brush free from honey. Some producers recommend the practice of blowing the bees from the combs to be extracted in preference to smoking. A jet of compressed air applied between the combs will take the bees to the box below without argument. Where an air compressor is not available, a carpet vacuum with a hose attached in reverse can be improvised and works very well. However, for most occasions the

smoking method seems to be the best. An empty super of combs placed under the box of honey to be taken provides a space where the bees can retreat. Otherwise, they will mill around in confusion.

Extracting Method 1: Uncapping

A new 20-gallon galvanized garbage can or similar container may be used in lew of a regular uncapping tank. Equip the container with a crossbar fastened across the top. Place a sharpened nail extended upward through the bar on which the frame will pivot for convenience in uncapping. Insert a square piece of quarter-inch mesh hardware cloth, shaped to fit the round can, with the corners turned down to keep it about 4 inches from the bottom, to facilitate drainage. At the bottom of this container a honey gate or drain pipe can be inserted.

Extracting Method 2: The Spoon Method

In the absence of an extractor the honey may be recovered by the use of a large spoon scraping the capped cells of honey. Leave a thin wall of the original foundation. Care should be taken not to puncture this thin wall, otherwise the bees may not rebuild a perfect comb. The frames can then be replaced in the super for the bees to rebuild the combs.

Capping honey is a delicious product. It is a universally-prized food item.

Honey obtained by the spoon method can be bottled

SPOON METHOD
OF EXTRACTING

POURING INTO CONTAINERS

for home use or for sale as chunk honey, or the wax can be separated from the honey by using a sun wax melter or placing it over low heat.

If you prefer extracted honey, a neighboring producer may, for a nominal fee, extract your honey.

CHAPTER 25

PREPARING FOR WINTER

Readying The Hives

In order to be prepared against any event it is safer to winter the bees in two-story hives. This is particularly advantageous during the spring build-up which allows the queen to maintain a broodnest of four to nine frames of brood. Since the brood chamber requires a temperature of about 90 degrees, outside combs are seldom used from brood rearing. Therefore, a single-story hive does not provide the desired space for the brood nest. A total of 40 pounds or more of sealed honey in a one or two story hive is safe winter storage in most cold northern sections. In southern sections and California where some honey is gathered by bees all year long, much less honey need be left. It is a good practice and vitally important in colder sections that the center frames in the upper story be full combs of honey. Should the bees decide to cluster in the upper story or in empty frames the result would be a dead colony, even though there is plenty of honey elsewhere in the hive. A lesson in unselfishness can be learned from these bees—although the cluster cannot move its position in zero-degree weather, there is a constant rotation of the outside bees moving toward the center with the well-filled bees relinquishing their places and moving to the cold outside circle.

In some northern sections winter packing is profitable. One method is to wrap black tar paper around the hive and over the top. Be sure to provide a top entrance (a 3/8'' hole will suffice), in case of deep snow or dead bees clogging the lower entrance. This arrangement will also provide for the winter cleansing flight.

The summer entrance should be made smaller to guard against mice entering, and also for warmth.

It is an interesting fact that under a normal honey flow, about 80% of the total amount of honey produced by the bees is required for their daily food, for producing their young, and for needed sorties for winter and spring. Bees can be wintered in a single ten-frame hive; however, in most cases it is profitable to maintain a two-story brood chamber. The amount of honey consumed during their semi-dormancy is negligible; but as spring approaches and brood rearing commences, the amount consumed is greatly increased. The exact amount is largely governed by the length and intensity of the winter. During a mild winter, more honey will be consumed because the bees are more active. A late cold spring, which deprives the bees of nectar from the fruit and other early blooms, will make a greater demand on the in-winter stores.

The hive inhabitants give off considerable moisture which can bring death by freezing within the hive unless provision is made for moisture to escape. The regulation inner cover provides for ventilation. A burlap sack placed underneath the top cover will absorb moisture. Another safeguard is to drill a 3/8" or 1/2" hole in the upper story above the lower entrance.

Winter Inspection

During the period of winter inactivity it is profitable and rewarding to visit your hive or apiary, checking for any outside disturbance, winter flight activity, etc. Don't be alarmed if you see forty or more dead bees in front of the hive. This is a normal condition caused from old bees being removed from the hive at the close of the season. On warm winter days the cluster will break sufficient to drag the dead bees from the hive. Mid-winter or early spring brood rearing will replace this loss.

Preparing For Spring

Winter is a good time to ascertain the needs for the coming season. If an increase is desired, the necessary hives and frames could be made up in the family workshop or purchased from a supply house. If the colony is to be increased, prepare one bottom box with a lid and at least two supers all equipped with frames wired and ready for foundation. If wired or plastic foundation is used it is not necessary to wire the frames.

Several factors will govern the amount of increase that could be realized during the next season. The United States is so varied in temperature, length of season, etc., that a blanket procedure cannot apply in all cases. Some sections will permit no increase, while others produce all increase but no surplus honey for the market.

In Emery County, Utah, I experimented with a hive placed in a sheltered location one-and-a-half miles from the nearest fields and three-and-a-half miles from a large alfalfa field down the canyon. The parent hive was increased five times during the same season. That season the five nuclei and the mother hive produced 210 lbs. of extracted honey plus sufficient stores for the winter.

CHAPTER 26

THE SECOND SEASON

Spring Inspections and Feedings

With the first warm days of spring you will need to inspect the bees. The flight activity from the hive will, in most cases, convey to you an indication of the conditions inside. If a colony has sustained a heavy loss in numbers (which could be due to the fact that they approached winter with a limited number of young bees) the bees flying from the hive will be few in number and will not show the vigor of a colony that experienced opposite conditions where brood-rearing was more substantial.

Since a temperature of about 90 degrees must be maintained in the brood chamber nest, it is not a good practice to open the hive for inside inspection during cold spring weather. The amount of winter stores still remaining can be determined by lifting the box for weight. The hive population, a warm open winter allowing greater activity, and other factors will determine the demand for honey stores.

If there is less than 30 lbs remaining then spring feeding may become necessary. The method of feeding will be determined largely by localities, weather conditions, feed availability etc. Syrup made from sugar is recommended only for spring feeding and is not advised for winter storage. Sugar lacks food elements essential for brood-rearing.

Outside feeding, in most cases, is seldom a good practice because of the loss of feed to neighboring bees. The bees become so ravenous and excited with the discovery of ready-made honey, they seem to forfit all sense of decency. They fight each other, drown in the syrup and when the supply is exhausted, they go in search of weaker hives, overcome the guards and take possession until every drop of honey has been carried away. This is called robbing. The unfortunate victims are either stung to death or left to starve.

A single colony isolated from other bees could be given outside feeding with more success. When a fast buildup of brood rearing is desired for the purpose of making increase or entering the honey flow season with a full force of adult workerbees, feeding is profitable even though ample stores remain in the hive.

Improved methods are currently being discovered for inside feeding. A method of feeding that will cause the least disturbance inside the hive especially through cold spring days is preferable. If hive stores are low and you have saved full combs of honey for spring feeding, then the feeding problem is readily solved.

Filling Empty Combs for Feeding

In the absence of extra combs of honey for feeding, or inside feeding equipment, try this method: place frames of built-out comb over a container and pour the syrup into the cells. The filled frames placed in position in the super will hold up for a short time by cappillary action. Therefore a loss of syrup will be sustained if the frames are not immediately placed on the hive to be fed. If honey is to be used for feeding, dilute it with fifty percent water and bring it to a boil. Sugar and water in the same proportions may also be used.

Artificial Swarming

Because of the variety of climates and conditions, several different methods of artificial swarming are practiced throughout the country. The beginner should consult a local beekeeper or current bee publications before attempting any of these methods.

This first method is designed particularly for the busy backyard beekeeper with limited spare time and experience. If, during a good honey flow your two story hive has eight or more frames of brood, you may wish to proceed with one of the following steps:

1. Place one half of the brood and honey into a new hive regardless of queen location. Two or three days later the queenless half will have started queen cells. If this queenless half is not on

the original location then switch places with the original hive. This will give it the support of the field bees in rearing the new queen.

2. When queen cells are in progress and the colony has eight or more frames of brood, they are ready to be divided. The first step is to find the queen. It will save time if you locate the queen early in the spring and place a drop of paint on her thorax, making identification easy. In searching for the queen, remove the combs gently and do not apply more smoke than necessary to keep the hive quiet. When the queen is found—and that will be a thrill— remove her (together with the bees and comb on which she is found) to the new hive box prepared for the increase. Add to her new hive one or more frames of brood and bees; then fill the remaining space with empty combs or foundation frames. Secure the lid and place her hive on the original stand, the original hive having been removed to one side a few feet away. The field bees from the old hive will return to their former location, thus giving the queen speedy support in building a new home. The young bees will remain in the original hive to take care of the unsealed brood and to complete the development of the queen cells. For a few days the hive will seem quiet and listless and without field activity. This first queen to hatch will proceed to destroy the queens in the remaining un-hatched cells. They will be pierced by her un-barbed stinger to re-move all competition and to comply with the law which calls for only one queen to reign in the hive. Then follow a few days of banqueting, being fed by the nurse bees, and a maiden flight which is a short honeymoon trip with the lucky drone (or we might say unlucky, for the price he pays for this one moment of thrill is his life). The fertilized queen then returns to her hive for a rest period and more feeding by the nurse bees. With the queen in full command, the hive now resumes the hum of activity. The young bees are now adults and go forth in search of nectar.

The queen now commences a two-or three-year period of productivity where 1000 or more eggs are laid each day. Can you imagine the number of eggs required to maintain a family of 50,000 to 100,000 or more at the peak of the honey flow, and this in view of the fact that her offspring live only about six weeks during the summer months, six months through the winter. The bee's wings, like a piece of cloth, do not rebuild with new cells. The collapse of the wing often occurs on the way home when the bee is heavily laden with pollen or honey or when it is facing strong winds.

Another method of artificial swarming is practiced in this manner: early in the spring the beekeeper encourages a fast buildup of hive strength by outside feeding of sugar or honey syrup (even though his bees have sufficient stores). The two-story hives must have at the swarming stage eight or more frames of brood with spring blossoms yielding nectar. Queen cells must not have been started; queens, previously ordered, should be on hand when the increase is to be made. The hive to be increased is treated thus: the old queen is removed to the lower story by smoking and brushing sufficient bees from the upper story (making sure the queen is below). Then place a queen excluder over the lower hive and replace the super (upper box). The following day the worker bees will have returned through the excluder to the brood above. The brood and bees above the queen excluder can then be transferred to a new hive and taken to a new location where a queen will be introduced.

When No Increase is Desired

Early in the spring before increased hive population, find the queen and clip one wing, rendering flight impossible. When the swarm ensues, the queen will crashland and be lost. The swarm will mill around for a while but then return to the parent hive. The loss of the queen will only delay swarming unless added room is provided.

If your bees swarm in spite of these precautions, you hive the swarm as usual and place it on top of the parent hive, but divided with a bottom board. Later on, remove the separating bottom board and cover and replace it with a single sheet of newspaper. As the bees chew holes in the paper they will become acquainted and united without fighting. A royal battle will determine which queen will reign. This guarantees a very strong colony for the honey flow.

Requeening

Bees are capable of requeening the hive, insuring the existance of the species over vast periods of time, living in hollow trees or clifts in rocks or ledges.

However, the introduction of improved queens will improve the temperament and add strength to their honey-gathering potential. A good practice is to requeen your present hive in the fall after the honey flow is over. This will insure ample brood rearing for winter and early spring.

Dead Bees

Should dead bees appear in great numbers in front of the hive during the honey flow, it is probably caused from poisonous insecticides sprays. If this continues the bees should be moved temporarily to a new location where they will hopefully find unsprayed sources of nectar. Home gardening sprays are a constant danger but generally the losses are not excessive.

Detecting and Treating American Foul Brood

Foul brood is a minute spore that develops when the bee is in the larva stage and completely destroys it, turning it into a dark glue-like substance. With a toothpick or match, try to remove the larva from the cell. It will string and smell very much like glue. The sealed brood will be darker in appearance. The capping will first show a pin-head hole and eventually be entirely dissolved. The final stage in the cell will be a dark dried up substance glued to the side of the cell.

Over the years this number one bee disease has been the source of tremendous expense and loss to honey producers, necessitating burning or boiling millions of hives. In more recent years the discovery of drugs has controlled the disease to a remarkable degree. However, every beekeeper great or small, must still exercise strict vigilance. Terramycin in the powdered form can be purchased from the bee supply houses. Sprinkle two or three teaspoonsful between the top bars or frames of the brood chamber near the back of the hive, thus preventing the terramycin from coming in direct contact with the brood. This will burn the brood. This operation should be repeated after ten days time. It is good practice to apply the drug to all colonies as a preventative, preferably during the springtime and after extracting. For fall treatment after the honey flow is over one application is considered sufficient.

Terramycin can be mixed with diluted honey or sugar water and fed through the regular feeding trays. The drug can be used without diluting but lesser amounts must be used (one teaspoon is sufficient).

Do not neglect this responsibility. It may save your bees and others. If American Foul Brood has developed, infecting one or more brood combs, the diseased frames can be removed and the colony given more frequent doses of the drug. However, when the disease has reached this stage, it is safer to boil the combs and start over. European Foul Brood is similar to American but less damaging and can be removed from the cell by the bees if the queen is removed from the hive.

Chalk Brood Disease

Chalk Brood Disease is a fungus that preys on the pupae, turning it into a chalklike substance which can be removed from the cell by the bees.

While not as severe as American Foul Brood it is a growing menace to be reckoned with. The U.S. Department of Agriculture has done considerable research work for the control of this disease.

Nosema

Nosema is a spore-forming protozoan, a muscular disease to which queens, worker bees and drones are susceptable. It shortens the life span of the queen and the worker bees, decreasing the normal range for foraging.

It is becoming widespread, especially offering a serious problem to package bee producers in some southern states and parts of California.

Over the past ten years the United States and Canada, in co-operation, have done extensive research work regarding the nature of Nosema and methods of control.

For other, lesser-known diseases, consult any leading Bee Journal, encylopedia, or your county or state bee inspector.

CHAPTER 27

GROUP BEEKEEPING

The preceding chapters of this book have dealt with the individual and his bees. Now we turn to some of the important advantages inherent in group beekeeping.

The Scouting Program

A scout troop might choose to maintain a few hives of bees for purposes of merit badge acquisition, fund raising and as an excellent educational activity.

Before beginning such a project, consider carefully the preceding portions of this book. It is very important that there be genuine interest and a willingness to become familiar with the rules of the hive. It would be wise to invite a neighborhood or commercial beekeeper to talk with the troop after some preliminary reading has been done. Even better would be an invitation to the boys and their leaders to put on some bee veils and observe the operations of the beekeeper. When the facts are in and the interest gauged, then proceed to secure equipment, choose a proper location, etc. Even though this might be a troop activity, direct responsibility should be vested in one or two boys to see that the project is sustained.

In one season all boys in the troop could earn the beekeeping merit badge and perhaps some money for troop activities also.

The Elementary School

Beekeeping would be a rewarding school project for some of the same reasons outlined in the preceding section. Again, the preparation for this experience cannot be over emphasized. The school library could maintain basic holdings cited in the bibliography of this book. Considerable reading, discussion, visits by guest resource people, and actual observations will set the stage for a successful project.

Practically every part of the school curriculum can be drawn into such a project. Look at the possibilities:

HISTORY and GEOGRAPHY—in researching the origin of bees, their locations and constraints imposed by varying climes.

MATHEMATICS and BUSINESS—in calculating hive demensions, costs, markets, profits.

NUTRITION—studying and testing the nutritional value of honey and experimenting with honey recipes in the cafeteria.

NATURAL SCIENCES, especially biology.

SOCIOLOGY—in comparing the intricate relationships within the hive with contemporary social systems and problems.

LANGUAGE ARTS—in reading, discussing and writing about the project.

ART and MUSIC—in painting and composing music related to bees.

Extracting equipment could be located in a corner of existing facilities in a workshop or even in a plant maintenance building. Space might also be provided for packaging the honey. If space cannot be secured for these purposes, one of the students might persuade his father to set aside a part of the garage or arrangements might be made with a local beekeeper.

The bees should be located at some safe distance within the playground properly screened, with a reasonably high fence in order to remove most of the hazard that might be involved if located in the mainstream of playground traffic. It is suggested that only a limited number of hives be used, perhaps three or four. Remember, it is important to locate them in the sun.

Colonies of bees kept on a school campus or in other public locations should be completely enclosed with at least an 8 ft. high panneling. A glass front to the south would permit observation and also admit the winter sun necessary for the bees in colder areas where winter cleaning flights are a must.

Bees will fly thru panel or board openings. A one inch mesh wire fence will not assure the flight pattern to be up and over.

A properly constructed enclosure will safeguard the public against bee attacks. A locked entrance will keep intruders out and will prevent wilful medeling which could stir the bees to anger, endangering the public and causing possible failure to the project.

Other Groups

Other beekeeping groups could be formed. Neighbors could participate together, or a bee colony could be a project for a large family or even elderly people in a convalescent home.

CHAPTER 28

A BRIEF HISTORY OF BEEKEEPING

Beekeeping dates back to the earliest world civilizations and has long been a part of both man's agricultural and religious heritage.

An article by James I. Hambleton, Division of Bee Culture Bureau of Entomology, U.S. Dept. of Agriculture observed that:

> If the statement seems too strong that the honeybee is more widely distributed than any other of our common insects, it can be said conservatively that the product of the bee is the most widely produced of man's food. Even such common foods as wheat and milk are not so universally known...
>
> In time they learned of the sweetness of honey which could be employed for many purposes. For centuries honey was the only sweet and it and beeswax were regarded so highly by the ancients that they wove into their religious ceremonies in one way or another frequent references to honey, wax and bees. Symbols representing various phases of bee husbandry are found in the earliest recorded histories. Man throughout his existence has been closely associated with the honeybee.
>
> Honey and beeswax were used in the payment of taxes and as indemnity. Conquered tribes and peoples paid off reparations in the form of honey and wax. To the present day beeswax plays an important part in the rites of the church.

In Mormon literature, for instance, the earliest mention of the honeybee is found in the records of the Jaredites, a band of people who left the Middle East at the time of the Tower of Babel. When they migrated to the American continent, they carried with them deseret, which interpreted is the honey bee. *(Book of Mormon, Ether 2:3)*. In another instance, Lehi, an ancient prophet who lived in Jerusalem with his family and friends during the reign of King Zedekiah, left Jerusalem and came to the American continent. Wild honey was recognized by them as a basic food. *(Book of Mormon, 1st Nephi 18:6)*

The following quotation from the U.S. Agriculture Handbook No. 335, briefly summarizes the history of beekeeping in the United States.

Beekeeping was unknown in the Western Hemisphere until after the first European settlers arrived. The natives of the West Indies used honey and waxes, which probably came from the native stingless bees (Meliponinae). Wax may also have been obtained as a plant derivative. The Indians of North America referred to honey bees as "the white man's fly," and regarded their presence as indicating the coming of white settlers....

The actual date of importation of the first colonies of honey bees (Apis mellifera L.) to North America is unknown. It was druing early colonization, since bees were important in the rural European economy. In 1622, bees were in Virginia and by 1648, beeswax and honey were abundant. In 1640, the town of Newbury, Mass., established a municipal apiary. (The expert put in charge became the town's pauper.) In 1641, bee colonies in New England were sold for 5 pounds apiece, the equivalent of 15 days labor by a skilled craftsman.

There is some question as to exactly when honey bees first reached the West, although the first importation of colonies that survived and multiplied apparently was in 1853. These were taken to San Francisco by boat from the Eastern United States. Extensive importations began shortly thereafter and the colonies prospered and increased rapidly. The mild climate and vast areas of flowers and shrubs highly attractive to bees offered seemingly unlimited beekeeping expansion. California soon became one of the most important beekeeping areas in the country.

In 1828, Moses Quinby took up beekeeping in New York; in 1837, the Reverend Lorenzo Langstroth acquired two hives in Andover Mass. These two men revolutionized the bee industry in the United States. Publication of their books in about 1852 signaled the advance to begin....

Langstroth's discovery of bee space led directly to the development of the modern hive. With the publication of his results, the U. S. Government Patent Office was deluged with designs for new hives. Every imaginable variation of Langstroth's design was attempted, many to avoid infringement on his patent. The fact that the patented feature was bee space, and not the frame itself, was ignored. Although Langstroth's patent was clearly valid, he had neither financial nor physical resources for the long court battles that followed, and he soon gave up the struggle to protect his patent. The deluge of hive designs ended; only a few sizes of hives are still in use today.

During this period the industrial revolution affected beekeeping, resulting in machine production of standard-sized interchangeable frame and hive. Two inventions, the wax comb foundation and the centrifugal honey extractor, completed the evolution of beekeeping from a tiny farming sideline to its present commercial status.

The idea of a centrifugal honey extractor came from an Australian, Major Hrushka. Prior to his invention, honey could only be removed by cutting the comb from the frames, crushing it and straining the honey from the wax. This was both time consuming and costly. Until the development of the extractor, the biggest seasonal honey crop for one beekeeper was obtained by Quinby and amounted to only 10 tons.

(Beekeeping in the United States, Agricultural Handbook No. 335, pages 2-3).

The Beehive State

Utah, although not the highest in honey production, is recognized for selecting the skep, an early man-made dome-shaped domicile for the honey-bee, as an important factor in designing the state emblem symbolizing industry.

UTAH TRAVEL COUNCIL

In 1848 five bee colonies brought by wagon train from the eastern states arrived in good condition. Walter E. Dodge in Feb. 1862 returned from California with five hives. The hum within voiced a protest over the long rough ride.

The early pioneers valued honey as a staple food and for medicine.

CHAPTER 29

BEE STORIES

The following are true experiences that occurred over the years as I followed my normal beekeeping pursuits. They are written for interest as well as to show the consequences when the rules of the bees are observed or disregarded.

The German Black Bee

My two sons, who live in the Bay Area of California, purchased eight colonies of bees to be divided for backyard beekeeping.

Mark placed his bees some distance away from his home on a terrace side hill. Kay has a half-acre lot landscaped with trees, grapevines and a 50-foot square set aside for the garden. His bees occupied a spot at the head of the garden as if they were to be guards; this they turned out to be. With no provocation, they would attack anyone who entered the fenced-in garden. Perhaps the more violent ones were a strain of the German Black Bee.

On our first visit to the bees, we took a comb of honey for family use. They became so angry they were intolerable, making it necessary to move them away. Had the bees been screened in with a board fence six to eight feet high, and with requeening from a gentler strain, they no doubt would have been successful.

In the case of Mark's bees tucked away on the side hill and seldom visited (or worked), they still persisted in attacking at a great distance from the hive. His colony also contained a strain of the black bees.

The German black bee has no place in backyard beekeeping.

Esther Dickey Becomes a Keeper of the Bees

Since honey is one of the basic foods listed in "Passport to Survival," and is now becoming a critical item, my daughter Esther (the author of the book) thought it right and proper that she should contribute in a small way and thus practice her own precept, "Be prepared and fear not."

The decision was made. We walked over the gently sloping three-acre plot, with the Dickey home created to overlook the beautiful countryside, in search of a location for her first hive of bees. After several locations were considered, Esther replied, "Dad, I cannot see my bees tucked away in some obscure corner, when

I would have to walk through the mud to watch them. I want my bees to be close by and part of the family. Could I have them on our front patio?" I replied, "Esther, that is the most unique location for backyard beekeeping in all my experience." We hastened out to the patio. The exact spot selected was directly under the large window of their double office where she and her husband Russell could observe with great interest and learn from nature the first lessons in the magic of the hive.

Does it sound ridiculous to you that bees be kept on the front patio? Would the bees take possession, keep away all visitors and even drive away the family? Most would say this was a crack-pot idea. Nevertheless, it can be done if the game is played by the rules of the bees.

Let us consider: the patio 14 x 18 feet was completely enclosed with a 7 foot high board fence. The plank floor, three dogwood trees and miniature pond which was part of the rock fountain were beautiful and offered safety from bee stings. The flight of the bee would be over the fence, completely out of the path of people. The well-fastened gate would offer security from intruders and small children.

What a rich experience it was for Esther and Russell to watch a stream of worker bees, about 70,000 in number, lifting into the air and flying off to the fields for nectar from daylight until dark, bringing home the sweetest of foods—80% for the use of the hive and 20% as a contribution for family use and storage.

With a swarm box prepared and nestled in the corner by the dogwood tree, how interesting it was to watch a swarm come forth from the parent hive and of their own accord, fly into the swarm box. Of course they might have chosen to cluster elsewhere, but it was worth a try.

Was this a good location for the bees? Let's consider it. It provided reflection of the sun during the winter months and protection from the rainy season of Oregon. It offered a place of warmth and dryness which was most essential, since moisture ordinarily develops within the hive and sometimes becomes a problem unless air circulation is provided. (By way of explanation, a small hole in the super is made at the top under the lid.) This location would be most beneficial to the bees during the beginning of brood rearing, adding warmth and shelter from cold winds.

With these advantages the swarming season could be advanced two or three weeks and the hive strength build to 75,000 or 100,000 strong, ready for the first honeyflow.

Since the Italian strain predominates, there was one more good reason the venture would be a success. The miniature pond was to be converted and used for the growing of watercress. The fresh running water for growing watercress would be very convenient for the bees and of great value, especially in the spring at the time of brood rearing. A composition of royal jelly cannot be made without water. The pollen in the cells as well as granulated honey must be mixed with water before it can be used.

Now the dreaming was over. Where and how would we obtain a hive of bees? A telephone call was made to a backyard beekeeper 12 miles away and the reply came, "I have no bees for sale except to you, Esther, and this will be in partial payment with my gratitude for all the quarts of milk and the food I have consumed in your home." Loren Ricks, once a shiftless boy without a father's guidance, is now a prosperous man with a family. The truck was made ready, the destination reached, a two-story hive heavy with honey was selected, loaded and ready for the new home. The price was discussed with these terms, "Esther, I will give you the honey and the bees without charge. You may return the hive or a replacement. The deal is closed!" The bees were now a part of the Dickey family. With interest we all awaited the outcome of this bold adventure.

Interest from Esther's window increased as the bees resumed activity from semi-hibernation during early spring and then summer days. At first they were flying in short circles with no apparent purpose except for the exercise of wings.

With the advance of spring and warmer days the flight pattern was straight over the high board fence and off to the fields and garden. With the same precision they soon returned some with armlets of yellow pollen neatly balled on their legs, some with their honey sacks filled with the first nectar of the season.

Little Avey, a granddaughter, wanted a closer observation. With her Grandpa as a chaperone she timidly entered the patio. Standing by the side of the hive her keen eyesight observed in great detail the way of the bees. If a guard bee came near, Avey would duck for protection, but the bees were not resisting the visit and apparently the family hive could be left on the front patio without danger if properly managed.

Interest climaxed with excitement when the bees swarmed. They rushed forth in disorder as if the hive were on fire and then circled within the patio until the sky was darkened with 50,000 or more pilgrims. When the queen came forth the circling mass lifted from the patio and, led by the scouts, clustered in a nearby bush where they were captured and returned to become the second colony on the patio.

The Italian's Experience With Bees
(An unusual story of natural swarming)

In the first years of my commercial beekeeping with a partner, William Parker, my son Mark (then about six years old) was employed as a helper. It was his assignment to carry the smoker and apply it when the hives were open for inspection, when empty supers of combs were set in place, and more especially when the filled combs were to be taken for extracting.

Mark's mother had made a suitable veil with sleeves and gloves and he became a beekeeper. He was to be paid in stock. When we moved to Salt Lake from Emery, my bees were turned over to the care of J. Fleming Wakefield of Provo. Mark's bees were moved and placed at our home in Murray on a one-acre lot, which proved to be a mistake.

The nearby Murray Smelter cast poisonous fumes over the blossoms, killing most of Mark's bees. That summer all were gone but two colonies and to save the bees Mark and I decided to move them away from the smelter area. A site was secured east of Draper in a small clump of oak brush right near an orchard. We moved the two hives including the dead ones in the hopes we could refill these hives the next summer by artificial or natural swarming.

Because of my employment with the Mountain States Honey Producers Association as General Field Agent, I was away from home much of the time. It was not until mid-summer that we had an occasion to go out and look at the bees. To our surprise we found that every hive was filled again with swarms of bees and that they were completely filled with honey. We stood in wonderment as to how this could be, because as a rule swarms will not occupy an empty hive. But while we stood wondering, an Italian approached and with great excitement and swinging of arms, he related this story in broken English.

He explained to us that they had swarmed and that a huge swarm had settled in a nearby fruit tree. As he attempted to cultivate this orchard near the tree, his one-horse cultivator mounted with a little girl brushed so close to the bees that they attacked the horse. The horse took fright and ran away. The girl fell from the horse, luckily out of the path of the bouncing cultivator coming behind. The man stood helpless and then came as fast as he could to rescue the girl. But this is the way he described the bees, and as he did so, it was impossible for us to refrain from laughing. He said, "You know the som of a bee, he no stay in the hive; he maka the big bunch in the tree; he stinga the H__ out of us."

Well, we sobered and told him we were deeply sorry and that we would move the bees, but that it could not be done until we returned with empty combs and boxes, to give the bees more room. To move them in the heat of the summer, filled with honey, would be impossible without smothering the bees. As we made further inspection, we observed new combs of honey that contained about 10 pounds of honey—a beautiful comb all sealed over. I said to Mark, "Let's go down to the Italian family and make peace with them. I think maybe they would appreciate very much having this beautiful comb of honey."

We did so. The family was outside eating the evening meal, and when they saw that comb of honey that we presented to them, the children danced with glee. I'm sure that had we taken the family a $100.00 bill, it would not have pleased them as much as this comb of honey.

Returning to the site, we observed that the bees had been clustered in the tree for a number of days. They had built comb attached to the limb and had really made it their home. We concluded the empty hives had been filled with swarms from some other apiaries, since one of the hives had swarmed and settled in the tree, the remaining hive could not have re-filled the six dead ones.

This is another mystery of bee behavior, one that is hard to explain, because bees usually do not remain in a cluster in a tree longer than a day or two.

An Experience With Fermented Honey

The following experience may be a valuable lesson where honey is kept for extended storage. Fully ripe honey will never ferment, but honey which is extracted before fully ripened may ferment after the first year of storage, particularly in warm temperatures.

In about 1914 I helped organize the Emery County Bee Association, and later became its executive secretary and marketing agent. The beekeepers of Emery County had organized for the purpose of buying supplies and marketing honey.

On one occasion we had an experience with fermented honey which was caused from excessive moisture. This occurs when the honey is taken from the bees before they have properly ripened the honey. Ripening occurs by fanning excess moisture before sealing the honey. When honey is capped over to at least one-third of the surface of the comb, it will not ferment.

The beekeepers had delivered their honey to the Price Warehouse where it would be prepared for shipment when a suitable market was found. But because of poor market prices, the honey was stored in the warehouse until the next summer. At this time I succeeded in making a sale with the Sioux City Honey Company in Sioux City, Iowa. The honey had been stored in lots, each owner having a lot number containing the name, the number of cans, and the color or quality of the honey. This information at the time of shipment was forwarded to the packing company including a sample bottle of each lot of honey, giving the color and number of cans contained in that particular lot.

As samples were being taken from the various lots, I discovered in Lot 14 the signs of fermentation. The honey, of course, was hard but this lot showed signs of breaking down. The honey was softening with tiny bubbles and gave evidence that under the heat of the summer sun, the honey would break down and ferment. This was something to be concerned about because should that occur, the entire car lot could be rejected by the receiver. I advised the owner of the lot and he came to the warehouse to determine the condition of his honey. We sampled a number of his cans and he said that I was unduly concerned, that there was absolutely no danger, and that the honey was all right. But I still refused to accept his honey. He said, "I'll be responsible for my honey, and take whatever price or offer the buyers will give, because I know that the honey will not ferment."

We shipped the honey in a box car which was often sidetracked in the hot sun. When the buyers opened the car doors, they discovered some honey caps had blown up to the ceiling and the soft fermented honey was oozing out of the can and was smeared all over the car. Of course, I received a telegram to come and receive the honey. I wired back and followed with an explanation by letter, telling them that only this lot could possibly contain fermented honey; and if they would proceed to clean up the car and isolate the fermented cans of honey, we would accept the price they felt to give us for any fermented honey. They accepted this offer and proceeded to unload the car. This particular lot was white honey, and by heating it to a certain degree while still in the cans, they could stop fermentation. It could then be blended with other honey and actually improve the color of it. The price they offered for this lot was less than half of the original price, but the owner accepted the offer.

This could have been a serious experience had they decided to reject the honey, which they well could have done. All the beekeepers could have sustained a loss and the entire car condemned. The experience proved to be of great value to other beekeepers. It was the first and only experience we had with fermented honey.

The Laramie Experiment

How far will a bee fly for nectar and still be able to return to the hive?

Around the year 1930, the Extension Service at Laramie, a branch of the Agricultural Department at Washington D.C., made a test mainly to determine how far a bee will fly and return to the hive with nectar. A colony of bees was taken into the dry desert eight miles from the nearest flowering plant where the bees could obtain nectar. It was of great and surprising interest to learn that they did fly the eight miles and returned, depositing honey in the hive as evidence. Of course, this distance is too great for practical use, but it did establish a point.

My Own Test Hive

A few years earlier than this my partner, William Parker, and I determined to run a test mainly to prove the value of a proper wintering and sheltering place for the bees and also the value of shelter from the cold spring winds. Our bees were located in Emery, Utah at an elevation of 6,000 feet. The winters are chilly with many windy, cold spring days. The bees, in defiance of the cold, went in search of dandelion and other early blossoms. Many of the bees fell victims to the weather and never returned.

From our apiaries of 400 colonies located in the surrounding fields, we selected a two-story hive; the queen was young and the colony prosperous. We transported the colony down into Millers Canyon, about one-and-one-half miles from the nearest fields of Emery and about three-and-one-half miles farther down the canyon to a large alfalfa field. We selected a little side canyon, a short distance from the main highway, which was further secluded and protected by huge boulders and sand rocks that had fallen from the ledges. In this little cove we placed the hive for wintering. They came through in splendid shape, and when spring came they were busy gathering honey from the early mountain flowers and a little later from the serviceberry and squawberry bushes, while the bees in the valley were still restricted because of cold weather.

The test hive reached the swarming stage several weeks in advance of the rest of the bees in the valley. When we discovered we had about eight frames of hatching brood and about 15 queen cells in the process of being sealed over, we decided to extend our experiment by determining just how far we could go with artificial increase or through artificial swarming, and still leave the bees with sufficient hive strength and stores for the winter.

Five boxes were placed in a semi-circle around the parent hive. They were all filled with ready-made combs. In fact, in this operation, ready-made combs were furnished for the entire experiment. First the queen was found, which is no small task in such a populous colony. The queen was left in the original stand with about two combs of brood, and the remaining space was filled in with empty combs. Then we proceeded to give each one of the five hives a comb of brood, bees and a queen cell. In cases where the amount of brood was small in the combs, two combs were given. The hives were closed up by placing the lid on each hive, and the bees were left to develop and hatch new queens. Of course, the field bees that were taken to the new hives returned to their original hive, being oriented to their original stand. The young bees in the new hives cared for the brood and completed the queen cells. When the queen cell hatched and a queen was mated, then activity resumed in these young hives. After this we were free to give the queen additional brood and bees. (A virgin queen may not accept brood and bees.)

We returned about 10 days later, and the parent hive had rebuilt to the normal amount of brood and bees. So from the parent hive we carefully took a frame of brood and bees (making sure the queen was not on that comb) and gently shook a frame in front of each of the young hives. The young bees entered the hive and were accepted. The field bees returned to the original hive. We repeated this operation several times, thus adding strength to each one of the young hives. We were indeed surprised to discover at the end of the experiment that from the original hive, including the increase, we obtained 210 pounds of extracted honey, still leaving each of the hives with sufficient stores for winter. We were also greatly surprised to learn that during the summer and during the height of the honey flow, the bees were streaming down the canyon three and a half miles away. We had supposed they would naturally return to the nearest fields but they were going to the large alfalfa fields which must have produced an abundance of nectar. The average yeild of the colonies in the field was about sixty pounds that year.

I doubt if this experiment could be carried on today. The heyday of bumper honey crops is probably in the past. Pollution, the use of poisonous insecticides and spray, and commercial fertilizers used in the soils, all seem to have wrought great changes and certainly have lessened the flow of nectar in the blossoms. They have also destroyed many birds, bees and insects.

The Bartholomew Story

Does a bee make a single trip a day when carrying nectar from the flowers or do they make ten or 20 trips? A number of years ago Mr. Bartholomew from the Extension Service of the Department of Agriculture in Washington, D.C., came by invitation to Emery and spent several days with us, teaching us how to prepare a winter pack for bees in a cold climate. During a conversation on bee behavior, he claimed that bees make only one flight a day when carrying nectar.

He related it as follows: He had spent many days sitting by a beehive. In the morning he would powder the bees with a certain color as they left for the fields and then wait through the day to determine if those marked bees ever made a second trip out for honey. He observed that the bees never made a second trip.

This was hard for us to believe. We had observed in our area that summer showers revived the blossoms and greatly increased the flow of nectar, especially of the white sweet clover along the ditch bank. The bees would, for the next few days, greatly increase the amount of honey taken in. The honey just seemed to come in much faster as compared to the day before the refreshing rain. How could this be if the bees made only one trip a day? Later research has confirmed that they do indeed make several trips per day. Certainly it can be observed that when bees are carrying ready-made honey or syrup, they can make a trip every 20 minutes.

A year later, a number of colonies were taken to a new location. They contained brood in all stages, from one-day-old eggs to the hatching brood and bees, but no queen. It was intended that the bees would produce a queen. The hives were unplugged after being located—that is, the plug of the entrance was removed to allow the bees their freedom for the following morning. By accident one of the hives was left plugged. We did not return to this yard for about two weeks, giving the bees time to produce a queen and the queen to be mated. On our inspection to see if the hives had succeeded in producing a queen, we noticed with great concern the plugged-up hive, and we supposed they would all be smothered. But on further investigation we found they had discovered a small hole in the burlap cover under the lid, large enough for one bee to pass out or return. To our surprise this colony had developed a queen that was mated and was producing eggs. They had also re-filled the comb from the hatching brood equalling the amount in the hives which had the full entrance opening.

Could it be possible if the bees made only one trip that this hive competed with the hives with full entrance?

Note: Notwithstanding the foregoing incident, it is the author's opinion that the bee makes many trips a day for nectar.

Peter Faulkner's Introduction to Beekeeping

Commercial beekeeping was introduced to Emery County about 1920 by Thomas Chantry, William Parker and John Bassett, all of Iowa. Peter Faulkner of Ferron, Utah, became interested and wanted to learn the business in the hopes that someday he could become a beekeeper. Accordingly, arrangements were made with Mr. Chantry, who was there on routine inspection work, to let Peter learn by observation.

The routine work at the apiary usually consisted of first an inpsection for foul brood, hive strength, the quality of the queen, and the necessary amount of pollen. All this was interesting to Peter—so much so that he failed to observe the crawling bees on the comb that had been leaned up on the outside of the hive to give more room to work the rest of the comb in the hive. Chantry noticed this and warned Peter not to stand too close because bees might crawl up his pants leg. But Peter's interest was too great, and it was not long until he stepped so close that he came in con-tact with a frame of bees. Soon the bees were crawling up his pant leg and commenced operation from front and behind. Peter began to slap wherever he was stung and finally exclaimed, "H____, Mr. Chantry, my pants are plum full!"

At that Peter was gone, no more to be seen that day in the bee yard. He was not discouraged, and he forgave the bees their inflictions; he continued his search for learning in the hopes that someday he would keep the bees. In a short time he learned the rules of the bees and became a successful beekeeper.

Tracking Bees

Henry David Thoreau, America's great nineteenth century writer, philosopher, and naturalist, tells of his experience with bees in the following interesting narrative taken from his journal for February 13, 1852:

Talking with Rice this afternoon about the bees which I discovered the other day, he told me something about his bee-hunting. He and Pratt go out together once or twice a year. He takes a little tin box with a little refined sugar and water about the consistency of honey, or some honey in the comb, which comes up so high only in the box as to let the lid clear a bee's back, also some little bottles of paint—red, blue, white, etc.—and a compass properly prepared to line the bees with, the sights perhaps a foot apart. Then they ride off (this is in the fall) to some extensive wood, perhaps the west side of Sudbury. They go to some buckwheat field or a particular species of late goldenrod which especially the bees frequent at that season, and they are sure to find honey-bees enough. They catch one by putting the box under the blossoms and then covering him with the lid, at the same time cutting off the stalk of the flower. They then set down the box, and after a while raise the lid slightly to see if the bee is feeding: if so, they take off the lid, knowing that he will not fly away till he gets ready, and catch another; and so on till they get a sufficient number. Then they thrust sticks into their little paint bottles, and, with these, watching their opportunity, they give the bees each a spot of a particular color on his body—they spot him distinctly—and then, lying about a rod off, not to scare them, and watching them carefully all the while, they wait till one has filled his sac, and prepared to depart to his hive. They are careful to note whether he has a red or a blue jacket or what color. He rises up about ten feet and then begins to circle rapidly round and round with a a hum, sometimes a circle twenty feet in diameter before he has decided which way to steer, and then suddenly shoots off in a beeline to his hive. The hunters lie flat on their backs and watch him carefully all the while. If blue-jacket steers toward the open land where there are known to be hives, they forthwith leave out of the box all the blue-jackets, and move off a little and open the box in a new place to get rid of that family. And so they work till they come to a bee, red-jacket perhaps, that steers into the wood or swamp or in a direction to suit them. They take the point of compass exactly, and wait perhaps till red-jacket comes back, that they may ascertain his course more exactly, and also judge by the time it has taken for him to go and return, using their watches, how far off the nest is, though sometimes they are disappointed in their calculations, for it may take the [bee] more or less time to crawl into its nest, depending on its position in the tree. By the third trip he will commonly bring some of his companions. Our hunters then move forward a piece, from time to time letting out a bee to make sure of their course. After the bees have gone and come once, they generally steer straight to their nest at once without circling round first. Sometimes the hunters, having observed this course carefully on the compass, go round a quarter of a circle and, letting out another bee, observe the course from that point, knowing that where these two lines intersect must be the nest. Rice thinks that a beeline does not vary more than fifteen or twenty feet from a straight one going half a mile. They frequently trace the bees thus to their hives more than a mile.

He said that the last time he went out the wind was so strong that the bees made some leeway just as a bullet will, and he could not get the exact course to their hive. He has a hive of bees over in Sudbury, and every year he sows some buckwheat for them. He has visited this buckwheat when in blossom when there was more than one bee to every six inches square, and out of curiosity has caught a number of the bees and, letting them out successively, has calculated by the several courses they took whose hives they came from in almost every instance, though some had come more than two miles and others belonged to his own hive close by.

A Bee In The Window

The warm smell of honey on the kitchen stove lured a bee from the backyard through the open door. Sensing captivity, it winged to the south window and crashed against the pane. Noticed at first but then forgotten for hours, the bee plied the window with his head hard against the glass, flying up and down and then across, repeatedly falling to the window sill, until finally it lay motionless. Sometime later I noticed the plight of this intruder and, feeling some compassion, placed a drop of honey within close range. The fragrant smell of honey brought the first sign of life. It moved feebly toward the honey, piercing the drop of sweetness with its tongue. The suction pump went into action, and the bee's little antennae waved in seeming delight at this unexpected good fortune. The bee moved away but quickly returned for a second helping, repeating this several times until its little honey sac was well filled. Then it stood motionless as if waiting to gain full strength.

Taking the bee by one wing, I carefully carried it to the open door and with the other hand in position as a take-off board, released my friend. It quickly lifted into the air and flew towards its home, enjoying freedom once more.

The Parable of the Unwise Bee
By James E. Talmage

Sometimes I find myself under obligations of work requiring quiet and seclusion such as neither my comfortable office nor the cozy study at home insures. My favorite retreat is an upper room in the tower of a large building, well removed from the noise and confusion of the city streets. The room is somewhat difficult of access, and relatively secure against human intrusion. Therein I have spent many peaceful and busy hours with books and pen.

I am not always without visitors, however, especially in summertime; for, when I sit with windows open, flying insects occasionally find entrance and share the place with me. These self-invited guests are not unwelcome. Many a time I have laid down the pen and, forgetful of my theme, have watched with interest the activities of these winged visitants, with an after-thought that the time so spent had not been wasted, for, is it not true, that even a butterfly, a beetle, or a bee, may be a bearer of lessons to the receptive student?

A wild bee from the neighboring hills once flew into the room; and at intervals during an hour or more I caught the pleasing hum of its flight. The little creature realized that it was a prisoner, yet all its efforts to find the exit through the partly opened casement failed. I threw the window wide, and tried at first to guide and then to drive the bee to liberty and safety, knowing well that if left in the room it would die as other insects there entrapped had perished in the dry atmosphere of the enclosure. The more I tried to drive it out, the more determinedly did it oppose and resist my efforts. Its erstwhile peaceful hum developed into an angry roar, its darting flight became hostile and threatening.

Then it caught me off my guard and stung my hand—the hand that would have guided it to freedom. At last it alighted on a pendant attached to the ceiling, beyond my reach of help or injury. The sharp pain of its unkind sting aroused in me rather pity than anger. I knew the inevitable penalty of its mistaken opposition and defiance; and I had to leave the creature to its fate. Three days later I returned to the room and found the dried, lifeless body of the bee on the writing table. It had paid for its stubbornness with its life.

To the bee's short-sightedness and selfish misunderstanding I was a foe, a persistent persecutor, a mortal enemy bent on its destruction; while in truth I was its friend, offering it ransom of the life it had put in forfeit through its own error, striving to redeem it, in spite of itself, from the prison-house of death and restore it to the outer air of liberty.

Are we so much wiser than the bee that no analogy lies between its unwise course and our lives? We are prone to contend, sometimes with vehemence and anger, against the adversity which after all may be the manifestation of superior wisdom and loving care, directed against our temporary comfort for our permanent blessing. In the tribulations and sufferings of mortality there is a divine ministry which only the godless soul can wholly fail to discern. To many the loss of wealth has been a boon, a providential means of leading or driving them from the confines of selfish indulgence to the sunshine and the open, where boundless opportunity waits on effort. Disappointment, sorrow, and

affliction may be the expression of an all-wise Father's kindness.
Consider the lesson of the unwise bee!

"Trust in the Lord with all thine heart; and lean not unto thine own understanding.
In all thy ways acknowledge him, and he shall direct thy paths." *(Proverbs 3:5, 6.)*

(James E. Talmage, *Improvement Era* 65:817 [Nov. 1962].)

Author's Note: Had Dr. Talmage's knowledge of bee behavior been equal to his ability as a writer, the ill fated bee would have gained freedom. The hand would not have been stung.

CHAPTER 30

A FINAL NOTE

It may be of interest as well as some future benefit for readers to review conditions that have developed since the author decided to write this book. Sugar, honey, and other basic foods have reached dangerous price levels, posing a threat to the peace and security of society. Yet it is comforting to note that, through backyard beekeeping, honey can be available to more people than many other basic foods. It is relatively easy and inexpensively to produce and does not require complicated manufacturing processes. Nor would it be subject to crippling strikes, controls, or political intrigue.

Backyard beekeeping can be successful. It can result in a warm, friendly working relationship between the master and the honey bee. There is not a more opportune time than the present to demonstrate the way to simplicity in the choice and preparation of wholesome foods. Related to the production of honey, the planting of nectar-bearing trees, shrubs, and field crops would have a direct and positive impact on ecology. The same can be said for the selection of non-poisonous sprays and organic fertilization.

Backyard beekeeping can spark the neighborhood spirit that would contribute to the solution of some of the world's serious human relationship problems.

Over the years the beekeeper and his busy friends have suffered the stigma of isolation because of fear of a bee sting. By following correct methods of handling and control and remembering the rules of the bees, fear will be replaced with confidence. City ordinances prohibiting the keeping of bees can be modified.

If this limited guide takes its place among the many others that have been written to guide the backyard beekeeper into a profitable, pleasant activity, the author will feel amply repaid for his efforts.

Finally, the author recognizes that some may consider as extravagant claims made in support of backyard beekeeping, but a tiny acorn becomes a mighty tree. Man has been his own undoing, but, by the same token, he will be responsible for the restoration of all things until that millenial condition of peace and goodwill is realized.

APPENDIX I

SOURCES OF FURTHER INFORMATION

You may wish to subscribe to one of the leading bee journals such as: *American Bee Journal*, published by Dadant & Sons, Inc., Hamilton, Illinois, 62341, and *Gleanings in Bee Culture*, published by the A. I. Root Co., Medina, Ohio.

Other sources of information may be found in your local library. Literature can also be obtained by writing for free government bulletins such as *Beekeeping For Beginners*, (U. S. Dept. of Agriculture) and *Home and Garden Bulletin*, No. 158.

See or write your local honey and bee supply company for pamphlets and books including *Starting Right with Bees.* (The A. I. Root Co., P. O. Box 1153, San Antonio, Texas, 78204) and the Merit Badge Series Booklet *Beekeeping*, (Boy Scouts of America.)

APPENDIX II

SUGGESTED LETTER IN DEFENSE
OF KEEPING BEES WITHIN CITY LIMITS

The following is a suggested letter which could be sent to the agriculture department in defense of keeping bees within city limits.

> Address
> City, State, Zip
> Date

To the Agricultural Department:

Causes for loss of honey bees and the serious decline in honey production are best known to our agricultural departments and to horticulturists.

A bumper honey crop in most areas is a thing of the past. Extremely high prices are prohibitive to many who have come to recognize the value of honey as a nutritious basic food.

Neighborhood or backyard beekeeping seems to be the best answer to this crisis. Among other requirements necessary to achieve success in smaller apiaries would be the revision of city ordinances that do not allow apiaries.

Over the past years bees have been and are now kept in New York, Chicago and other large cities on penthouses, roof tops and even in attics of dwelling homes. They can be kept in neighborhood back lots without danger of bee stings to children or adults.

This can be done by proper handling so that the bees are not provoked to anger and by providing an enclosure requiring the bees to fly up and over, as well as safeguarding small children from approaching the hive in the absence of adults.

The honeybee has many points in its favor. In addition to honey production they rank first in pollinizing flowers, fruit tree blossoms and field crops. They are directly responsible for the stability and continuity of agriculture. They can provide adventure and merit badges in scouting, study and organization on the school campus, food and fun for the family, and relaxation and nature secrets for the hobbyist.

Only a single charge can be made against them—they sting. This can be controlled.

We invite your careful consideration and will greatly appreciate your support.

> Yours truly,

INDEX

A

American Foul Brood, 120

B

Bees

Adult Bees, 91

Bee Box-bottom, 80

Bee veil, 80; illustration, 95

Bee wire, use of, 80

Brood, description of, 80; nest, 80

Characteristics and habits, 85, 86, 87

Dead bees, 120; Development of, 90

Educational benefits of, 76

Getting started, 92

Beehive (measurements for) 80; Diagram of hive, 93; Beehive State, 126

Illustrations of Bees, worker, queen and drone, 89

Investment questions, 78, 79

Pollination benefits, 75

Preparing for winter, 113, 114

Rules for working with bees, 103, 104, 105

Stings, what to do, 105

Swarm description of, 84; How to capture, 96, 97, 98

Young bees, activities of, 90

Bee Stories

"The German Black Bee" 127

"Esther Dickey Becomes a Keeper of Bees" 127

"Italians Experience with Bees" 130

"Experience with Fermented Honey" 131, 133

"The Laramie Experiment" 133

"My Own Test Hive" 133, 135

"The Bartholomew Story" 135

"Peter Faulkner's Introduction To Beekeeping" 136

"Tracking Bees" by Henry David Thoreau, 136

"A Bee in the Window" 138

"The Parable of the Unwise Bee" by James Talmage, 139

Utah, Beehive State, 126

Beekeeping

Backyard, 75; group beekeeping, scouting program, 122; in school, 122; in neighborhoods, 123; letter in defense of backyard beekeeping, 145

History of, 124, 125

Transfer of Bees, 96, 99, 100, 101

C

Calcium, 14

Carob, nutritional value of, 17

Cell, description of, 80

Chunk-honey, what is, 81

Colony, description, 89; activity of colony, 89

Comb, definition, 81

D

Dextrose, 13, 14

Diseases of bees, 121

Foul Brood, 81; treatment of, 120

Chalk Brood, 121

Nosema, 121

E

Embedder, use of, 81

Enclosure for bees, purpose and how to build, 100

Equipment needed, 92

European Foul Brood, 121

Extracting honey, 110; Method 1, 111; Method 2, 111

F

Feeder, description and use, 81

Feeding, methods, 117

Filtering, process, 81

Foul Brood, definition 81; detecting and treatment of, 120

Foul Brood, European, 121

Foundation, description, 81

G

Gear, protective clothing for the beekeeper, 95

German bee, 93